T0201559

THE SCRAMBLE FOR THE POLES

THE SCRAMBLE FOR THE POLES

THE GEOPOLITICS OF THE ARCTIC AND ANTARCTIC

KLAUS DODDS AND MARK NUTTALL

polity

First published in 2016 by Polity Press

Polity Press
65 Bridge Street
Cambridge CB2 1UR, UK

Polity Press
350 Main Street
Malden, MA 02148, USA

ISBN-13: 978-0-7456-5244-3
ISBN-13: 978-0-7456-5245-0(pb)

A catalogue record for this book is available from the British Library.

Library of Congress Cataloging-in-Publication Data

Dodds, Klaus.
    The scramble for the poles : the geopolitics of the Arctic and Antarctic / Klaus Dodds, Mark Nuttall.
        pages cm
    Includes bibliographical references and index.
    ISBN 978-0-7456-5244-3 (hardcover : alk. paper) -- ISBN 0-7456-5244-1 (hardcover : alk. paper) -- ISBN 978-0-7456-5245-0 (pbk. : alk. paper) -- ISBN 0-7456-5245-X (pbk. : alk. paper) 1. Geopolitics--Polar regions. 2. Polar regions--International status.
I. Nuttall, Mark. II. Title.
    G593.D 64 2015
    320.1'20911--dc23
                                                                                   2015009805

Typeset in 10.5 on 12pt Sabon by
Servis Filmsetting Ltd, Stockport, Cheshire
Printed and bound in the UK by Clays Ltd, St Ives plc

For further information on Polity, visit our website:
politybooks.com

CONTENTS

FIGURES

PREFACE

When we think of polar scrambles we might be forgiven for alighting upon comparisons with historical, and indeed contemporary, scrambles for Africa and other parts of the colonized world. In the nineteenth century, European states including Britain and France were at the forefront of carving out territories and digging up resources as their domestic societies underwent further industrial and urban development. Some scholars detect a 'new scramble for Africa', this time involving not only the United States but also actors associated with the global South, such as China, India and Brazil (Carmody 2011). The role of China in particular has elicited the greatest attention, including its so-called 'no questions asked' policy, as relating to the governance and human rights cultures of client states in Africa. But states are not the only actor involved in this scrambling enterprise. International institutions such as the World Bank and the International Monetary Fund, alongside a ragbag of agents including multinational corporations, criminal cartels and private military contractors, are also complicit in the 'opening up' of African economies and the associated patterns of dispossession, exploitation and violence (Watts 2012).

The historical and contemporary polar scrambles we interrogate in this book are not without their own stories of dispossession, exploitation, marginalization and violence. As Edward Said (1978, 1993) reminded us, alongside a coterie of geographers and historians of the colonial past and present, the way in which land, people and environments are mapped, administered and exploited is rarely free from contestation. We find, for example, plenty of evidence of what Joseph Conrad once termed 'militant geography', a temporal and spatial marker of what we might think of as predatory forms of mapping and charting, ever eager to enclose yet more territories and resources (Driver 2000; Gregory 2004). This lust for knowledge and geography, informed and enriched by a series of imaginary geographies, positioned the colonial world as awaiting the civilizational imprint of European explorers, missionaries, scientists and military personnel.

In the nineteenth century, there was no shortage of 'militant geography' in the Arctic region, as British and other European explorers and sailors sought to transit through the Northwest Passage. If moisture and heat acted as deterrents to those militant geographers in Africa, the cold and ice were equally capable of exposing the limits of European bodies, ideas and practices. While moisture and heat remain unforgiving on bodies and objects, frigid polar waters and arid land-based ecosystems end up preserving the 'colonial present'. The Arctic landscape is littered with the traces of earlier European exploitation and administration in the form of huts, settlements, tracks, whalebones and mining projects.

On 9 September 2014, Canadian Prime Minister Stephen Harper was photographed standing in front of a large television screen. Unveiling an image of the hull of a sunken ship, he pointed his finger at the screen with obvious excitement. The image in question was that of HMS *Erebus*, one of the ships belonging to the ill-fated Franklin expedition, a Victorian-era adventure designed to chart and open up the Northwest Passage as a shorter route to Asia for the benefit of European maritime travel and trade. What followed was an interesting, even remarkable, series of associations being offered up by Harper. In essence, this shipwreck was being positioned as a proxy for Canadian Arctic sovereignty even though the expedition was organized, financed and promoted by the British government and scholarly societies, such as the Royal Geographical Society in London, years before the creation of the Canadian federation. Sir John Franklin, the leader of the expedition in 1845, might have been surprised to be told that almost 170 years later a Canadian political leader would appropriate his ill-fated journey and lost ships for national purposes.

But this expedition was unusual (figure 0.1). The disappearance of the Franklin expedition in 1848 provoked impassioned appeals by Lady Jane Franklin and extraordinary searches, resulting in the further mapping of the North American Arctic (David 2000; Potter 2007). Franklin left England in May 1845 aboard HMS *Erebus* alongside HMS *Terror*, commanded by Captain Francis Crozier. Sailing up the west coast of Greenland into Lancaster Sound, the ships made progress through Barrow Strait and circumnavigated

Figure 0.1 The Search for Sir John Franklin (reproduced with permission of the British Library)

Cornwallis Island. They spent the winter south-west of Devon Island at Beechey Island (where three crew members died). In September 1846 HMSs *Erebus* and *Terror* were stranded in pack ice somewhere north-west of King William Island and were abandoned in April 1848. The British Admiralty sent a search mission in 1849 commanded by James Clark Ross, but no sign of the ships was found. Over the next decade, government-sponsored and privately financed expeditions (including some sent by Franklin's wife Lady Jane) were dispatched. The Franklin search became something of a national obsession, as many have written about it in depth, both in terms of the Franklin expedition itself and the search for the Northwest Passage (e.g. G. Williams 2003, 2009; Lambert 2009), and in a broader thematic treatment of nineteenth-century Arctic exploration, polar imaginaries and imperial Britain (e.g. Spufford 1997; David 2000; Hill 2008). Traces of men and equipment were found by some of the search expeditions, as well as by Inuit, including notes detailing events up to April 1848 (Franklin is known to have died in June 1847). John Rae was involved in several search parties and in 1853–4 he ventured north to map parts of Canada's continental coastline. He met Inuit at Repulse Bay who sold him some objects, including silverware, from the Franklin expedition. The Inuit reported they had seen white men dragging sledges, and that they had died of starvation, and finally that they had discovered bodies that showed evidence of cannibalism (Spufford 1997). Rae returned to England in 1854 with the expedition's artefacts and the news that some of the desperate and starving survivors might have resorted to cannibalism. These revelations were met with shock and disbelief and offended the sensibilities of those who believed 'civilized' British officers and crewmen would not have begun to eat one another (including Charles Dickens who felt compelled to express his abhorrence in an essay). The Inuit testimony was dismissed as the false account of 'savages', yet led Lady Franklin to believe that some of the expedition survivors might be living among them. The lost ships remained elusive, however, and became central to an enduring narrative about the Arctic. As Glyn Williams put it, 'No episode in the history of oceanic enterprise offers a greater contrast between anticipation and disillusionment than the centuries-long search for the northwest passage' (Williams 2009: xv), and the Franklin episode is a supreme example of this.

Franklin's expedition – and the many voyages that ventured north to look for him and the crews of HMSs *Erebus* and *Terror* – also figured prominently in the nurturing and development of a Canadian northern identity. The Northwest Passage, remote, distant, yet symbolic of heroic deeds, polar exploration and of 'the true North strong and free' has long provided inspiration for writers, poets, folk singers, other musicians and playwrights (Grace 2001). Few Canadians are likely to visit these northern waters, yet the historical adventures, epic journeys and disasters associated with the discovery and mapping of Canadian national territory in the Arctic and the emergence of Canada as a nation have become inextricably linked to and bound up with contemporary ideas of sovereignty and Canada's aspirations to become an Arctic power. In 1992, Parks Canada, a Canadian national heritage authority, declared HMSs *Erebus* and *Terror* to be historical monuments even though no one knew their exact location. The ascribing of such a status was testament to the sense of the powerful presence of the ships and their association with the idea of Canada as a northern place. In some ways, the feeling that the Canadian North is suffused, even haunted, with the essence of Franklin's two ships has been important to the story Canada has been telling itself and others about its identity and its place in the circumpolar world. Finding the remains of one of them has not solved the mysteries of the Franklin expedition but it has been useful for Canadian political discourses about sovereignty and claims to Arctic territory and resources. As the Canadian novelist Margaret Atwood (1991) reflected, the North remains a fertile imaginative landscape, filled with 'strange things' and strangers.

Over the years, various branches of the Canadian government, oil companies such as Dome Petroleum and private salvage hunters have been drawn to the enigma of the missing ships. In the most recent search, which led to the discovery of HMS *Erebus*, Shell Canada was involved, raising intriguing possibilities of why a multinational energy company might be concerned with such a quest in the first place. One factor, of course, might be that Shell Canada owns a large number of drilling leases in the North American maritime Arctic and perhaps senior executives thought that being associated with the search for the wreckage of an imperial British Arctic expedition was a good thing. Better to be seen as having an interest in polar heritage than being associated with global climate change and rapacious resource extraction in the Arctic region and beyond. It is worth recalling at this point that Shell has suspended its drilling operations on a number of occasions in the North American Arctic, and has faced considerable criticism from environmental groups and northern communities for its current and planned activities. Recently the Danish toymaker Lego was criticized for its long-standing relationship with Shell.

On the other hand, as scholars such as Adriana Craciun (2012, 2014) have noted, the connection between nineteenth-century expeditions, resource extraction and what we might think of as geopolitics and security, have a historical and gendered provenance that is difficult to ignore. Ever since the first European encounters with Arctic regions, explorers, traders, sailors and scholars have all played their part in promoting the interests of various states and companies in mapping, exploiting, administering and controlling spaces such as the Northwest Passage and the Arctic Ocean. Such attention to the Arctic and the efforts expended in sending expeditions of discovery, claiming ownership, asserting sovereignty and exploring the region's resource potential furnished the inspiration for Jules Verne's 1889 novel *The Purchase of the North Pole*, in which Barbicane and Company 'announced that it had "acquired" the territory for the purpose of working – "the coal-fields at the North Pole"!' at an auction in New York. A Canadian Prime Minister of European descent (with the support of agencies such as the Royal Canadian Geographical Society) is in that sense part of a longer trajectory of agents, objects, ideas and practices eager to find extractive value in Arctic ecosystems, even if it was just passing through them rather than hunting seals and whales, fishing, cutting timber and drilling for oil.

The discovery of Franklin's vessel should be one of those moments when we pause and think about those aforementioned historic connections and contemporary framings of the Arctic as a space for projecting a specific national sovereignty and identity politics. Harper claimed that the Franklin expedition was helping to 'map together the history of our country'. Running counter to that nationalist appropriation is another important aspect of the 'discovery story'. Inuit oral histories and traditional knowledge played a crucial role in transmitting memories and stories about the expedition and, despite the earlier ignorance about the power of indigenous testimony, such stores of knowledge were acknowledged by Parks Canada to be crucial to the latest search operations. All of which raises an awkward juxtaposition; the invocation of a long-lost nineteenth-century British expedition as a lynchpin to Canada's self-identity while at the same time acknowledging in the Arctic context the long-standing presence of Inuit remains the most evocative expression of Canadian sovereignty. For a Canadian Prime Minister deeply committed to promoting the idea that Canada is a 'northern nation', the discovery of a long-lost ship quickly became caught up in a highly opportunistic campaign to promote, once again, Canadian Arctic sovereignty.

For a geographer and an anthropologist, by way of contrast, this shipwreck news story, as we have just noted, is indicative of how Arctic geopolitics works in the here and now, including the constellations of power, knowledge and geography that make possible scrambles past and present. Past associations, albeit selected with great care by politicians and the like, are combined with contemporary opportunism and an outlook towards the future that is at times fearful, and at times hopeful. For politicians, the submerged wreck becomes both an object of Arctic geopolitics and a site for Arctic geopolitics. It is one that is more hopeful than, say, concentrating on past episodes of forced relocation of Inuit or contemporary concerns about poverty, housing and domestic violence. The point about associations is that powerful agents of Arctic geopolitics, such as prime ministers and presidents, pick and choose where possible. As the head of Pauktuutit, the organization representing Inuit women in Canada, Rebecca Kudloo noted at the time, 'If the [Canadian] government is willing to spend millions of dollars on a missing Franklin ship, why aren't they spending millions of dollars on violence against women?' And as British-based historian of nineteenth-century Arctic exploration, Shane McCorristine concluded, 'The remains of Franklin himself are still missing and the *Terror* is still lost but in a curious way, which I think the Canadian government recognizes, this expedition remains a haunting presence in the Arctic, a ghost story that continues to fascinate' (2014: 100).

Prime Minister Harper wanted Canadian personnel to discover the wreck and its location in the Northwest Passage because this is highly significant for a country that still worries about the mobility of others in a maritime space that it considers part of its 'historic waters' rather than an international strait (Steinberg 2014). Franklin's dream of an accessible Northwest Passage seems to be closer to realization as polar sea ice appears to be melting away. Others dispute that historic waters designation and believe that the passage is just that – a place for third parties to transit through without being impeded either by the Canadian government and/or sea ice (Byers 2013). Thus, the shipwreck, as a previously lost object with a Victorian English provenance, is actively enrolled in a more contemporary geopolitical project regarding security, sovereignty and stewardship.

Our Franklin vista is, thus, intended to open up a broader conceptual and empirical landscape involving both the Arctic and Antarctic. In what follows we outline and evaluate a series of entry-points for making sense of the contemporary 'scrambles' and 'scrambling' affecting the Polar Regions. We use these terms guardedly but do so because they are commonplace in media, academic and political literatures, and reportage. Moreover, we believe that they help us to make sense of what is at stake – politically and geographically. There are a multiplicity of 'scrambles'; scrambles to gather geographical knowledge about the seabed, scrambles to fish the Southern Ocean, and scrambles to 'open up', 'save' and/or 'protect' the Arctic and Antarctic. There are 'drivers' that are empowering these 'scrambles' and we conclude in our last chapter that these 'scrambles' carry with them a series of demands whether it be to speed up, slow down, intensify, refrain or block. Such 'scrambles', we posit, do not mean that the Arctic and Antarctic are doomed to be conflict-laden spaces in the future. Rather, we believe that these scrambles carry with them multiple futures, some of them more hopeful instead of dreadful, and some of them more likely rather than simply possible.

Throughout, we think of scrambles in two senses – scramble in the sense of preparing to act (often associated with war-like gestures such as scrambling jets) but also in the sense of ideas and things being broken up or scrambled like a radio broadcast that becomes unintelligible to the listener. Geophysical and geopolitical change is afoot. The Arctic is warming, the Antarctic is warming and cooling depending on where you care to investigate, and most observers would acknowledge that interest from the wider world in both Polar Regions is far greater than it was say in the 1940s and 1950s. It is also more complex, involving states with historical interests in the Polar Regions and states and non-state organizations who have more recent, even tangential, associations. In February 2015, a suspected case of illegal fishing in the Southern Ocean involved a New Zealand warship attempting to apprehend fishing vessels belonging to a Spanish syndicate with ships registered in Equatorial Guinea. Old clichés such as the Antarctic and Arctic being 'poles apart' just won't do anymore (if they ever did, given global trading networks and the movement of fish, seal and whale products as well as minerals from the Polar Regions to the rest of the world).

Our interest in polar scrambling encouraged us to travel, live and reflect on and in the Arctic and Antarctic. Our biggest debts of gratitude must go to our respective families in London (Klaus Dodds) and Edmonton (Mark Nuttall) for allowing us to make those journeys north and south and for spending long periods away from home. Klaus Dodds is immensely grateful to colleagues at Royal Holloway and elsewhere for many conversations and collaborations with Peter Adey, Duncan Depledge, Simon Dalby, Stuart Elden, Alan Hemmings, Valur Ingimundarson, Timo Koivurova, Alasdair Pinkerton, Richard Powell and Phil Steinberg. An ESRC Research Seminars Grant (with Richard Powell, 2010–12) proved invaluable for bringing together critical polar scholars and led to an editorial collection entitled *Polar Geopolitics*. The Geopolitics and Security Group at Royal Holloway, University of London remains a great place to be part of and the masters students who participated in his polar studies option are thanked for their engagement. He also thanks those who have supported his Arctic-based research, including the British Academy, the Canada-UK Council, the House of Lords Select Committee on the Arctic and the Royal Norwegian Embassy in London. Mark Nuttall thanks the Department of Anthropology at the University of Alberta, and the Greenland Climate Research Centre, Greenland Institute of Natural Resources and Ilisimatusarfik/University of Greenland in Nuuk for institutional support and for continuing to give the time, space and resources to carry out research on a number of polar topics, as well as colleagues and students who are too numerous to mention, for discussion, conversation and reflection. He is grateful to research grant support from the Henry Marshall Tory Chair Research Programme at the University of Alberta, from the Academy of Finland, and from the Greenland Climate Research Centre. We both

PREFACE

gratefully acknowledge the support of a British Academy-Association of Commonwealth University Joint Project Grant (2010–11), which enabled us to begin some initial planning for this book. Our subsequent planning took shape during conversations in Edmonton, London and Nuuk, and the chapters that follow are expansions and reflections on a number of themes that have preoccupied us since. We thank the two anonymous reviewers for their careful reading and very useful criticism, and we owe a debt of gratitude to Rachael Squire for reading through the manuscript and offering incisive comments throughout. Ian Tuttle did a splendid job copy-editing the manuscript. None of the aforementioned colleagues and institutions bears any responsibility for the analysis presented in this book.

We would like to thank the following for permission to reproduce images in this book: Figure 0.1 is reproduced with permission of the British Library; Figure 3.1 is reproduced with permission of *Popular Science*; Figure 4.1 is reproduced with permission of the British Antarctic Survey; Figure 5.2 is reproduced with the permission of the Government of Greenland; Figures 6.1 and 6.2 are reproduced with permission of Jean de Pomereu; Figure 7.1 is reproduced with permission of Greenpeace and Nick Cobbing.

We owe sincere thanks, lastly, to Louise Knight, Pascal Porcheron and colleagues at Polity for their patience and goodwill. This book has been a long time in the making.

— 1 —

SCRAMBLING FOR THE EXTRAORDINARY

In 1868, the Arctic explorer and physician Isaac Israel Hayes delivered a lecture to the American Geographical and Statistical Society entitled 'The progress of Arctic discovery', where he noted the onset of 'great scrambles' to acquire geographical knowledge about new territories, usually for commercial and political benefit (Hayes 1868). He was not alone in imagining the North Pole and the prospect of an 'open polar sea' as a powerful incentive for further scrambling, as nations sent men in ships, planes, balloons and airships or on skis in the hope of discovering new geographical points and commercially appealing spaces (Craciun 2009).

For much of the eighteenth century onwards, explorers, traders, administrators and scientists sought to map, to colonize and to administer the North American Arctic. It was a scramble for territory and for the resources that lay at and below the surface. We can also see attempts to find and transect the Northwest Passage or the Northeast Passage as forms of scrambling to get *through* and *across* the Arctic, not so much for the importance of discovery and knowledge of the Arctic regions in and of themselves, but for global imperial ambitions and the expansion of colonial ventures (Bloom 1993; Bloom et al. 2008).

Hayes was also not alone in advocating further effort to map and chart the Arctic region. In his 1860 account of Arctic exploration and discovery, Samuel Smucker underscored the importance for Great Britain to acquire a greater knowledge of polar geography and northern maritime routes for the country's 'vast and yearly increasing dominion, covering almost every region of the habitable globe', and he argued that it becomes necessary that she should keep pace with the progress of colonization, by enlarging, wherever possible, her maritime discoveries, completing and verifying our nautical surveys, improving her meteorological researches, opening up new and speedier periodical pathways over the oceans which were formerly traversed with so much danger, doubt, and difficulty, and maintaining her superiority as the greatest of maritime nations, by sustaining that high and distinguished rank for naval eminence which has ever attached to the British name. (Smucker 1860: 34)

From the early part of the nineteenth century, demands were placed on treasury chests and sponsors to fund, equip and dispatch polar expeditions, and on ships and men to sail, crew and map in northern regions for the purpose of reaching and extending power and influence over other places, peoples, ecologies and items of trade. For Smucker, the chart of Britain's colonies was 'a chart of the world in outline, sweeping the globe and touching every shore' (Smucker 1860: 34), but he was dismissive of the enduring importance of a north-west passage to Asia and pragmatic about its use. If it was ever to be found, he said, it would always be a hazardous and protracted journey to get through it and navigate the ice-choked waters – the fact that the polar seas of the northern hemisphere were thickly clustered with various lands as well as ice only convinced him of their impenetrability. The Northwest Passage would only be a useful sea route – and therefore the arduous nature of exploration, and the sufferings and perils of Arctic voyaging worth enduring – until shorter and quicker routes to Asia were made possible 'by railroads through America, or canals across the Isthmus' (Smucker 1860: 33).

At the time Smucker was compiling his compendium of discoveries in northern regions, others were hard at work assessing the Arctic's resource potential. Henrik Rink reported on the surveying and state of knowledge of Greenland's minerals, prospects for extraction and 'mining-speculation of private companies' (Rink 1974 [1877]: 79), while the Yukon Gold Rush in the late 1890s saw thousands of hopeful fortune-seekers scrambling, literally, up and along the Chilkoot Pass and other routes to the Klondike (Berton 1972; Porsild 1998). Charles Mair's *Through the Mackenzie Basin*, his 1908 account of the signing of Treaty 8 in the Athabasca district north of Edmonton, celebrated the possibilities of access to the great resource potential of northern Canada.

While such 'scrambles' predated the coinage of the term 'geopolitics' in 1899, it was precisely those kind of power-knowledge scrambles by colonial powers and post-revolutionary republics like

2

the United States that inspired an interest in how state power, literally, rested on mapping, moving and exploiting the earth (Ó Tuathail 1996). For the earliest geopolitical writers, the ebb and flow of European empires in Africa, Asia and other parts of the world was a source of fascination and even fear, as contemporaries such as British geographer Halford Mackinder worried about the prospect of future conflict over a world that was ever more mapped, colonized and exploited (Dodds and Atkinson 2000). The Polar Regions were not immune from this auditing, and resources such as minerals, fish, seals and whales were caught up in transnational and inter-imperial rivalries, with indigenous peoples, energy corporations and communities becoming enmeshed in those global resource scrambles (Anderson et al. 2009).

Making sense of historical and contemporary scrambles over the Arctic and Antarctic requires us to discuss three elements. The first element involves some reflection on how the Arctic and Antarctic have been defined and delimited in the past and present. As Hayes sensed in his 1868 lecture, where the Arctic began and ended was a moot point. Did sea ice signify a biogeographical boundary or was the presence of open water indicative of an Arctic that might be as much sea as it was ice and snow? This then allows us to move onto addressing both scrambles and scrambling, i.e. as object and as verb. Exactly what was, and is, being scrambled over, and who should be involved in those scrambles? Finally, we conclude with a sense of the geopolitical consequences of scrambling with some examples that will prepare the reader for a more detailed examination of how the Polar Regions have been caught up in power-knowledge scrambles, and made complicit in scrambles *inter alia* to secure access to resources, to generate knowledge, to exert power over peoples and societies, and to claim and administer territories, both onshore and offshore.

Defining the Arctic and Antarctic

The manner in which we define places such as the Polar Regions is variable. As critical geopolitical scholars have noted, our very definitions are always deeply geopolitical, highlighting some spaces, objects, relationships and communities at the expense of others (Dalby 1991; Ó Tuathail and Dalby 1998; Dodds 2012). For example, although people may think they know what the Arctic is and where to locate it on a map (as well as what it is supposed to look like), defining and delineating its southern boundaries are tasks that embroil researchers

3

in controversy. Definitions of the Arctic vary considerably according to scientific, environmental, geographical, political and cultural approaches, perspectives and biases (which are reflected in the different ways the working groups of the Arctic Council define 'the Arctic'). To complicate this further still, climate change is eroding many of the physical boundaries and features, and reshaping the contours of geography that have been drawn as seemingly fixed points on maps, such as the tree line, the southern extent of discontinuous permafrost, ice shelves, glaciers and the distribution of perennial sea ice. The presence of ice has long been thought of as precipitous for the mobility of indigenous peoples and more latterly European explorers in the Arctic region, while its projected disappearance and potential absence intimate prospects for increased access and new shipping routes.

In February 1968 British adventurer and polar traveller Wally Herbert stepped out from Point Barrow, Alaska, onto the sea ice with three companions. For the next 15 months they travelled by dog team and sledge across the Arctic Ocean to Phipps Island, north of Spitsbergen. Their sledges were designed so that they could be converted to boats for crossing stretches of open water. They moved slowly with their dogs pulling their equipment across the permanent pack ice or rowed from floe to floe, across an ocean strewn with the rubble of floating ice. At the end of this long journey, they had not only made a claim to become the first people to cross the Arctic Ocean, they may have become the first to reach the North Pole without using any form of motorized transport, given that Robert Peary's claim to have done so in 1909 remains controversial. In terms of adventuring, Herbert and his colleagues probably made a last great polar journey in the sense of achieving something that had not been done previously, but they also made the kind of crossing of the Arctic Ocean, primarily on ice, that may no longer be possible because of geophysical changes involving sea ice, ocean and wind currents, and subsurface, surface and air temperatures. Yet icescapes of continual flux and a process of topographical reshaping are part of the lived experiences of many who live in the Arctic, and who perceive the environment as one of emergence and becoming (Nuttall 2009).

So, where does the Arctic begin? And how far south does it extend? (See figure 1.1). We could say something similar for Antarctica as well (figure 1.2). For example, how far does the Antarctic region extend? And where do oceanic bodies such as the Southern Ocean begin and end? These are not the questions of the geographical pedant, rather they carry considerable potential for so many areas of interest to us. The boundaries that we impose on the earth's surface,

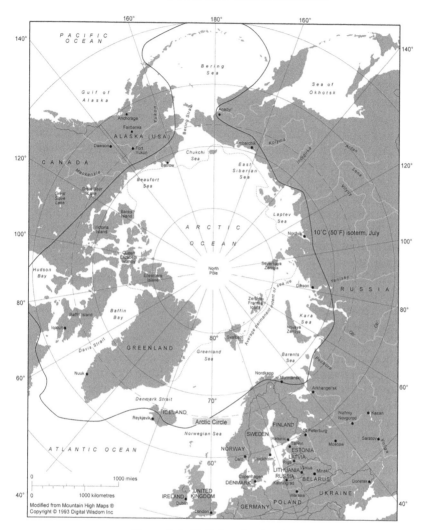

Figure 1.1 Defining the Arctic

depth and height influence and shape our legal systems, our governance, our resource management, our strategic awareness and our
cultural imagination. In the Polar Regions, where human habitation
has been demanding and prone to be humbled by immense physical
and environmental challenges and constraints, these boundaries and
demarcations can take on a quixotic quality at times. But they remain
important nonetheless.

5

Examining the Arctic, initially, many physical scientists would say that the biogeographical definition criteria of the region must include high latitude, long winters and short, cool summers, low precipitation, glaciers, ice sheets, permafrost, frozen lakes, rivers and sea in winter, and the relative absence of trees. But science is not a single discipline. To practise science means to seek knowledge about the world, and scientific disciplines have different views of how to go about it. An astronomer, for instance, would suggest the southern boundary of the Arctic could be established as the latitude beyond which the sun does not set at high summer, or rise during the depths of winter. This occurs at that imaginary line called the Arctic Circle at 66°33'3N. Where the Arctic Circle is supposed to lie is determined by the angle of the earth's axis in relation to the plane of its motion around the sun. This inclination means the sun's rays never shine immediately straight down on the Earth's surface north of the Arctic Circle and cannot effectively warm it significantly. This has the effect of reducing the amount of solar heat that the Earth's surface can absorb at high latitudes, creating what are arguably the most recognized Arctic characteristics – long, cold and dark winters, and short, cool summers with constant daylight for several weeks.

This astronomical determination does not influence how oceanographers define the Arctic, however. To them, the Arctic is the region where ocean temperature remains near the freezing point of salt water (about 1.7°C) and its salt content about 32 parts per thousand. A terrestrial ecologist, on the other hand, might describe the Arctic as existing only beyond the tree line, i.e. the point beyond which it is not possible for trees to grow. These astronomical and physical characteristics make the Arctic seemingly easy to define for those in specific fields, and to fix on maps and globes, yet Arctic-like conditions are found far south of the Arctic Circle and many Inuit, regarded often stereotypically as a quintessentially Arctic people dwelling far to the north, live in parts of Canada, Greenland and Alaska which are several hundred miles south of the Arctic Circle. If the tree line is taken as the southern boundary, western Alaska, the Aleutians and Iceland would be considered Arctic, although according to strict climatic criteria they would be excluded (Nuttall and Callaghan 2000).

According to climatic definitions, the Arctic is the region north of the 10°C isotherm. An isotherm (meaning a line of equal temperature) is a line on a climatic map linking points with the same mean annual temperature. The 10°C isotherm marks the southern limits of the high latitudes of the planet where the average monthly temperature is at or below 10°C, and the average for the coldest month is below 0°C.

The 10° isotherm more or less follows the tree line but it does not correspond to the Arctic Circle. Both the 10°C isotherm and the tree line, however, may diverge as much as 100 km in some areas, and both extend beyond 70°N in Norway and as far south as 55°N in Canada's Hudson Bay region. Sweden and Finland extend above the Arctic Circle, but these countries lie south of the tree line and the 10°C isotherm. The 10°C isotherm also makes a loop around the Bering Sea as far south as the western tip of the Aleutian Islands. Within these limits, Arctic conditions prevail over northern Fennoscandia (Nuttall and Callaghan 2000).

A closer look at various regions of the circumpolar North contradicts the stereotype of the Arctic as a land of perpetual cold, snow and ice, and it is worth noting that the anthropologist and explorer Vilhjalmur Stefansson pointed out some of the inconsistencies and difficulties of locating the Arctic in two versions of a chapter he called 'The North that Never Was', published in *The Friendly Arctic* and *The Northward Course of Empire* in the early 1920s. It is a place with incredible biodiversity and cultural diversity. The Arctic climate varies significantly depending on the location and season and is more accurately described as a collection of regional climates with different ecological and physical characteristics. Mean annual temperatures vary greatly. Despite popular assumptions, the North Pole is not the coldest place in the Arctic because the ocean moderates its climate. The coldest temperature in the Northern Hemisphere, −70°C, was recorded in the Oimyakon Basin in north-eastern Siberia, not far north of the Arctic Circle. Fairbanks, Alaska, which is south of the Arctic Circle, often records lower temperatures than Barrow, Alaska, which is far to the north of it on the Beaufort Sea coast. In terms of average temperature, Canada's coldest city is often Yellowknife, which is situated below the Arctic Circle, but the cities of Edmonton and Winnipeg much further south can also experience winter temperatures in the −40s. Often during a Canadian winter, Edmonton is colder than Whitehorse, Yellowknife and Iqaluit on account of northern Alberta's subarctic continental climate. Edmonton is also often considerably colder than many other places in the circumpolar world, which are located much further north. Because Greenland's capital Nuuk is warmed by the Gulf Stream, its temperature on a typical day in January may be −5° to −10°C, while residents in Edmonton shiver at −30°C. On the other hand, the southward extension of cold air masses plunge Edmonton into a deeply-frozen winter lasting several months. Yet despite this, although many people may agree that Edmonton is a northern city, they would not consider it to

7

be in the Arctic despite being on the same latitude as the southern-most fringes of Nunavut that extend into James Bay and further north than the southern coast of Labrador.

In the case of the Antarctic, the region is frequently defined and delineated with reference to latitude, climatic characteristics, ecological qualities, political and legal boundaries, as well as through appeals to its sublime wilderness and endangerment. The Antarctic as an area is sometimes defined as referring to the ice, rock and water below the Antarctic Circle. The Antarctic Circle is distinguished from Antarctica, which refers to the landmass that is the southern polar continent. The Antarctic Circle (defined as 66°S of the Equator) like the Arctic Circle (defined as 66°N of the Equator) experiences at least one day of continuous daylight every year (the December solstice), and a corresponding period of continuous night time at least once per year (the June solstice).

When it comes to the maritime governance of the Antarctic, the Antarctic Convergence is a powerful political and geographical marker but is not a latitudinal delineation. Sometimes called the 'polar frontal zone', it is a term used to describe a zone of transition in which the cold waters of the Southern Ocean meet the warmer waters of the Atlantic, Indian and Pacific Oceans. The convergence itself varies from year to year, depending on sea temperature and climatic trends. So these are flows that make, remake and unmake a zone of some 30–50 km in width, encircling the polar continent, and stretching north of South Georgia and Bouvet Island. The convergence roughly coincides with the mean February isotherm (10°C) and lies around 58°S, considerably north of the Antarctic Circle. Air and sea surface temperatures change markedly once one crosses the Antarctic Convergence. The Antarctic Convergence is significant because of the wealth of marine life, especially plankton and shrimp-like krill – the food of choice of birds, fish and whales – that is found there. As such, it also means that fishing stocks and sea birds tend to be concentrated around the Antarctic Convergence, leading to greater interest in managing these areas of the Southern Ocean as opposed to others.

The Antarctic Convergence is not equivalent to the Southern Ocean. While the Southern Ocean is often defined as being south of 60°S latitude, there is still disagreement as to whether it might extend further north. The British naval explorer Captain James Cook used the term to describe the vast seas of 50°S and below, but the International Hydrographic Organization (IHO) established the northern boundary at 60°S in 2000. Interestingly, the IHO had previously been unwilling to define the Southern Ocean in recognition that there was a lack of consensus on the mater. For example, many Australians and New Zealanders regard the water off the cities of Adelaide and Invercargill as the start of the Southern Ocean, thus consolidating their sense of these southerly cities as 'Antarctic gateways'. The IHO's usage of the 60°S latitude for its tentative definition of the Southern Ocean coincides with how the Antarctic is defined in the 1959 Antarctic Treaty. This treaty is the most important source of governance in the Antarctic region and Article VI of the Treaty notes:

The provisions of the present Treaty shall apply to the area south of 60° South Latitude, including all ice shelves, but nothing in the present Treaty shall prejudice or in any way affect the rights, or the exercise of the rights, of any State under international law with regard to the high seas within that area.

This area of application entered into force in June 1961, and over 50 signatories including the United States, Russia and China accept its fundamental provisions including its legal definition of the Antarctic (figure 1.2).

These lines and zones we have discussed are not exhaustive but each, in their own way, intervenes in how the Arctic and Antarctic are understood and impacted upon. In a more imaginative sense, we might acknowledge appeals to the sublime and wilderness. For nineteenth- and twentieth-century artists, explorers and scientists, the Antarctic was as much traced via the sublime as it was tentatively mapped and charted. As an artistic and literary expression, this notion refers to the capacity of things in nature to overwhelm the human mind by their sheer grandeur and immense possibility. A place or landscape might, as a consequence, inspire awe or provoke terror. So the sublime refers to something beyond the calculable and measurable, and more to a state of mind. The Antarctic in this particular sense is a true frontier of the human mind and body, and a testing ground of men and their capabilities in particular. Apsley Cherry-Garrard's memoir of Robert Falcon Scott's last expedition, *The Worst Journey in the World* (1922), memorably referred to the Antarctic as a place of privation and suffering. As he noted caustically, 'Polar exploration is at once the cleanest and most isolated way of having a bad time which has been devised.' But it could also be compelling, he felt, remarking enthusiastically, 'And I tell you, if you have the desire for knowledge and the power to give it physical expression, go out and explore.'

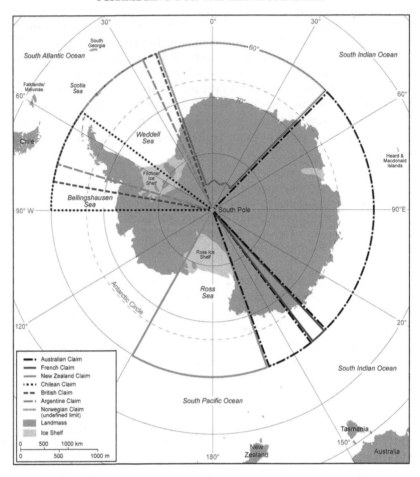

Figure 1.2 Defining the Antarctic

In the Arctic, nineteenth-century artists such as Edwin Landseer and Briton Riviere were indicative of this interest in the sublime as part of a wider enchantment with the northerly latitudes in Euro-American societies – immense icebergs, atmospheric phenomena such as the Northern Lights, polar bears and the vulnerability of ships and sailors attempting to traverse through the Arctic filled their work. Housed within the Founders Building of Royal Holloway, University of London, Landseer's painting *Man Proposes, God Disposes* reflects on the aftermath of the loss of Sir John Franklin's last expedition. Landseer's painting, exhibited at the Royal Academy in London in 1864, purported to represent the Arctic as a space where polar bears

10

would ravage the remains of humans consumed by the vagaries of the Arctic weather and sea ice. The allegations surrounding the expedition that suggested the last survivors might have engaged in cannibalism perhaps make the polar bears analogous to those men desperately seeking food as their circumstances deteriorated. While the painting was controversial with the families of the dead expedition members, including Lady Franklin, this fascination with Arctic landscapes and fauna such as the polar bear is a reminder that there might be more imaginative boundaries at play in terms of how the Polar Regions have been thought of as embodying the monstrous, the wild, the sublime and the exceptional.

We did not begin to write this book wedded to particular definitions of the Arctic and Antarctic, whether via lines of latitude, average summer temperature and/or tree line, and/or the sublime. But what we are interested in is the lines we draw and/or impose on maps, especially when used to define national territory and authorize sovereign authority over thousands of square miles of continental land, islands, ice and water, as well as people, societies and cultures. Although they are often assumed to be regions worthy of comparative study, the Arctic and Antarctic are polar opposites in several important ways. They are fundamentally different geographically in that a large portion of the Arctic consists of the ice-covered Arctic Ocean, which is surrounded by many islands and archipelagos, and the northern parts of the mainland areas of the North American and Eurasian continents, whereas the Antarctic is an ice-covered landmass surrounded by an ocean. They also have different environmental patterns, climatic systems and wildlife habitats – for instance, there is much lower terrestrial biodiversity in the Antarctic than in the Arctic – and there are no indigenous peoples living in Antarctica. Antarctica has also been subject to an international framework for environmental management and conservation under the Antarctic Treaty System for the past 50 years, whereas no such regime exists in the northern circumpolar region. The Arctic region comprises the northern parts of eight nation states, whereas, while the international community does not recognize sovereignty by states over any portion of Antarctica, seven countries have made territorial claims to parts of the continent.

Of course, one of the fundamental differences between the Arctic and Antarctic is the enduring presence of indigenous people and more recent settlers in the Arctic region compared with the transient population of scientists and other personnel working in Antarctica for national polar programmes. Indigenous peoples of the Arctic include the Iñupiat, Yup'iit, Alutiit, Aleuts and Athapaskans of Alaska; the

11

Inuit, Inuvialuit, Athapaskans and Dene of northern Canada; the Kalaallit and Inughuit of Greenland; the Saami of Fennoscandia and Russia's Kola peninsula; and the Chukchi, Even, Evenk, Nenets, Nivkhi and Yukaghir of the Russian Far North and Siberia. We do not set out to discuss any one particular society or culture in any detail, but we are attentive to ethnographic and indigenous accounts and articulations of how Arctic peoples have depended for thousands of years on the living resources of land and sea, as hunters, fishers and reindeer herders. Today, many indigenous communities across the Arctic continue to depend to a considerable extent on harvesting and using living terrestrial, marine and freshwater resources. The most commonly hunted species are marine mammals such as seals, walrus, narwhals, beluga, fin and minke whales and polar bears; and land mammals such as caribou, reindeer and musk-ox; and fish such as cod, salmon, Arctic char, redfish, northern pike, capelin and several other species are caught and consumed. Reindeer herding is widespread across Siberia and northern Fennoscandia, while sheep farming provides the basis for local livelihoods in parts of southern Greenland. Animals are a source of food, with the skins of marine mammals and furs from caribou and reindeer providing material for clothing and other products, but they also figure prominently in the cash economies of local households and communities (Nuttall 1992; Caulfield 2000; Dahl 2000; Kalland and Sejersen 2005).

Indigenous peoples maintain a strong connection to the Arctic environment through these activities, which provide the basis for food production, in a way that marks them out from many non-indigenous communities. One defining attribute of being indigenous is that it refers to the quality of a specific people relating their identity to a particular area and to their traditional cultural and economic dependence on local resources. Fishing and whaling communities and sheep farming families in Iceland, Faroes and Norway, and Finnish reindeer herders remind us, however, of the richness of place-based societies that makes the circumpolar North far more diverse as a lived, human world than the tendency for many studies to focus attention largely on indigenous culture reveals. Yet, despite the continued reliance on living resources, the northern indigenous world is increasingly cosmopolitan and diverse and, in many communities, the legacies and experiences of colonialism, cultural disruption and rapid social change are lived realities. Since the 1970s, there have been significant political changes in the Arctic that have included land claims in Alaska and Canada and the formation of regional governments in Greenland and Nunavut. Settlements include the Alaska Native Claims Settlement

Act, or ANCSA (1971), Greenland Home Rule in 1979 (followed 30 years later by Self-Rule), the James Bay and Northern Quebec Agreement (1975–7), the Inuvialuit Final Agreement (1984), and the Nunavut Agreement of 1992 (the territory of Nunavut was inaugurated in 1999). These political changes have often included changes in the ways that living and non-living resources are managed and, in some cases, have given indigenous people control over some institutions of governance as well as local and regional economies. A greater degree of local involvement in resource use management decisions has been introduced, including, in some parts of the North (more so in Canada than Russia, for example), the actual transfer of decision-making authority to the local community or regional level, thus enabling the use and application of indigenous knowledge in matters of governance.

In addition, significant steps have been taken with innovative co-management regimes that allow for the sharing of responsibility for resource management between indigenous and other uses and the state. As self-government is about being able to practise autonomy, the devolution of authority and the introduction of co-management often create the conditions that allow indigenous peoples opportunities to improve the degree to which management and the regulation of resource use considers and incorporates indigenous views and traditional resource use systems, but often with varying degrees of success (Huntington 1992; Nadasdy 2003).

Nonetheless, and despite these differences between the two Polar Regions, the influence of the Arctic and Antarctic over the function of the Earth system is emphasized in the scientific literature. They are both sites of considerable international political interest and intense scientific activity, as well as ecologically-sensitive regions that conservationists work to protect and environmentalists campaign to keep free from development. For example, an increasing number of non-Arctic states are expressing interest in both the resource potential and the governance of the Arctic Ocean, with some states and non-state actors such as conservation organizations like Greenpeace and the World Wide Fund for Nature (WWF) calling for the Arctic to be 'saved' from high seas fishing and oil and gas exploration through moratoriums and legally binding commitments to oversee the management of the region and its resource development (with some campaigning for a prohibition on oil and gas exploitation in ice-covered areas). Both regions are also attracting large numbers of tourists eager to experience what are represented as some of the world's last wilderness areas and the presence of large cruise ships provokes anxiety

13

amongst environmentalists over the impacts increasing numbers of visitors have on the environment and wildlife habitat. The loss of tourist ships such as the MV *Explorer* in the Antarctic Peninsula region in November 2007 heightened anxieties about search and rescue and possible loss of life. The sinking of MV *Explorer* was considered one of the better possible scenarios in terms of maritime accidents because at least everyone was rescued even though the ship was lost (Stewart and Draper 2008). In those frigid conditions, the outcome could have been a lot worse.

The Antarctic and Arctic are governed by particular and often very specific institutional arrangements and international initiatives of conservation or protection that are being tested by the transformative effects of unprecedented environmental change, as well as broader and far-reaching social, economic and political global processes. Climate change, pollution, ocean acidification and resource extraction bring new pressures and challenges to the Polar Regions, as do broader shifts in global architectures where states such as China and South Korea are understood by Western observers to be 'rising'. As we discuss in detail in chapter 4, the negotiation of governance arrangements, whether they emerge in the form of either the Antarctic Treaty, the Arctic Council, or indigenous and regional political settlements, brings into sharp relief particular spatial understandings of and connections constituting the Arctic and Antarctic. Those connections are never politically innocent and any exercise of power, control and dependency creates geographies of inequality, division and marginalization – the by-product, one might say, of such spatial settlements.

In the remainder of this chapter, we outline the way in which we have set out to encounter the contemporary geopolitics of the Arctic and Antarctic through a scrambling foray. Drawing inspiration from the interdisciplinary literature of critical geopolitics, history of science and technology, feminist studies and, what we might term, critical polar studies (itself informed by post-colonial and gender theorizing), we place emphasis on the manner in which these regions are both represented but also subject to a range of performances and practices, including what is a mixture of statecraft and stagecraft (e.g., Naylor et al. 2008; Nuttall 2008; Kafarowski 2009; Saarinen 2009; Glasberg 2013; Doel et al. 2014; Pinkerton and Benwell 2014). It is difficult to write about the Polar Regions, for instance, without giving due acknowledgement to the kinds of activities that so often accompany 'icy geopolitics' – we simply note the worldwide impact that the planting of a Russian flag on the bottom of the central

Arctic Ocean in August 2007 engendered (Dodds 2010; Gerhardt et al. 2010; Powell 2010). As Patricia Seed pointed out with reference to the Columbian encounter (i.e. European colonialism from 1492 onwards), ceremonies of possession matter and they persist into the contemporary era (Seed 1995). The use of a term such as 'engender' might also usefully remind us that it is men that so often embrace these frontier performances, whether they be explorers, scientists, soldiers and/or political leaders.

Scrambles and Scrambling the Arctic and Antarctic

The story about a shipwreck and its discovery in contested waters, as we noted in the preface, offers a powerful insight into how the Polar Regions are routinely framed as exceptional spaces and extraordinary places, which appear to demand and/or encourage extraordinary actions by states and other parties, including corporations, individuals and even entire communities. Millions of Canadian dollars have been spent on the hunt for the Franklin ships – HMS *Erebus* was found on the eighth attempt to locate the vessels – and one remains to be discovered. But the search for such relics is only one element in what we have termed 'scrambles and scrambling', including the juxtaposition between expeditions (including the wrecks of past ones) and the legal-technical and geopolitical processes involved in trying to pin down and make sense of claims to sovereignty and security.

Around the time of the Russian flag-planting expedition in 2007, for example, there was a scramble of sorts by journalists, academics and more popular writers to publish articles and books on the 'scramble for the poles', the 'rush for resources' and the 'new Great Game in the North'. Notable works include Alun Anderson's *After the Ice: Life, Death, and Geopolitics in the New Arctic* (2009), Roger Howard's *The Arctic Gold Rush: The New Race for Tomorrow's Natural Resources* (2009), Richard Sale and Eugene Potapov's *The Scramble for the Arctic: Ownership, Exploitation and Conflict in the Far North* (2010), and Charles Emmerson's *The Future History of the Arctic* (2010). These books deal largely with issues of development and resource extraction, and of sovereignty and territorial claims against a backdrop of climate change and rapid meltdown in the circumpolar North and the emergence of what has been called 'the new Arctic'. As earlier geopolitical writers would have recognized, the physical geographies of the earth (land, sea and ice) were invoked as exemplars of this scrambling story (Davis 1910).

15

While cautious of adopting uncritically a widely used framing device, the verb 'to scramble' is a helpful departure point for understanding the contemporary geopolitics of the Polar Regions. It draws attention to both the creative manner in which we understand places like the Arctic but also to something arguably more fundamental. As many scholars have noted, climate change is being felt most keenly in the Northern Hemisphere, and specifically the Arctic. Countless studies have warned not only of sea ice thinning but also of growing instability of ice sheets in, for instance, Greenland (see e.g. ACIA 2005; AMAP 2011). The consequences of such melting and thinning of ice and permafrost are still to be determined, but scientists are already providing us with a series of likely global climate change scenarios based on the consequences of sea level rise and changing water temperatures.

In the Arctic, the long-term trend is towards further warming with consequences for sea ice thickness and distribution, ice sheet stability and the condition of permafrost. Put simply, the Arctic Ocean may well be radically changed in the present century with some scenarios intimating that it will become largely free of perennial sea ice in the summer season in the next decade or so. Other studies have warned of intensifying acidification of the Arctic Ocean with attendant implications for marine life and for communities that depend on marine harvesting and tourism (AMAP 2014).

One consequence of these biophysical changes, especially in the Arctic, has been to bring to the fore a gamut of economic and geopolitical opportunities. But there may also be challenges, especially for the five Arctic Ocean coastal states of Canada, Denmark/Greenland, Norway, Russia and the United States. Indeed, the whole notion of a 'scramble' is enabled in large part by a sense in which the Arctic Ocean in particular is becoming more accessible to those five coastal states as well as outside agents and organizations. The shadow of new shipping routes, enhanced mineral extraction, and military projections of power and influence are routinely cited as evidence of such accessibility alongside other activities such as the 2007–9 International Polar Year, which provides a different kind of reminder of the Arctic's incorporation into global scientific networking. At its most extreme, nightmarish visions of an unruly and loosely governed icy realm (but one that is losing its ice) are unleashed, with an accompanying sense that governments have to be seen to be acting in the face of continued uncertainty.

Empowering such territorial imaginaries is the spectre of accelerating climate change and the prospect of intensifying resource-led

16

opportunities – leading to new kinds of geographical framings from a 'frozen desert' in the past to a new 'polar Mediterranean'. While such opportunities may well be over-stated, in terms of the immediacy of new shipping lanes and/or resource potential, their very articulation matters. The emphasis on future potential has resonance with stakeholders and it encourages a form of geopolitical and geo-economic boosterism, where some things, some people and some spaces get amplified and branded as 'inviting', 'opportunistic' and/or 'threatening'. The booster-like scripting of the Arctic Ocean, in particular, as a weak, disorganized and ungoverned space was particularly noticeable in 2007–8. Imagined as insecure, frontier-like and even dangerous, this characterization, however ignorant of existing international maritime law and governance structures such as the Arctic Council (English 2013), was used by national leaders such as Stephen Harper of Canada and Vladimir Putin of Russia to propose new interventions and investments ranging from constructing icebreakers to installing new research stations and port facilities.

In the Canadian context, moreover, the idea that the Arctic might be more accessible, and thus open to other parties, provides a powerful geopolitical context for legitimating state and non-state practices. When Harper tours the Canadian Arctic, as he does every summer, he does so in large part to demonstrate *his* and *his* country's resolve to defend national sovereignty, especially in the High Arctic Archipelago region. Being seen by Canadians and others is a critical element in that touring, especially if the idea is to challenge any perception that the Canadian portion of the Arctic Ocean is a weakly governed space. Harper is always eager to be photographed standing next to members of the Canadian armed forces and surrounded by large Canadian flags.

The scrambling of the Arctic has arguably contributed to a series of state and non-state led practices and subjectivities empowered by a strong sense of what has been termed by Matthew Hannah (2010) as a form of actionism (sic). A form of geopolitics where due emphasis is given to action rather than considered reflection, especially when confronting the apparent spectre of 'weak governance'. The deployment of 'sovereignty and patrolling exercises' such as Operation Nanook, an annual military-based exercise in northern Canada, becomes all the more understandable. In the latest incarnations, the exercises involved Canadian, Danish and US military service personnel and local indigenous communities in a series of scenarios including the possible environmental consequences of an oil spill in the Canadian Northwest Passage. The exercise emerges, at least in the Canadian context, as a cipher for strong governance. Secure polar space, under this geographical imagination, is well integrated with southern communities and territories. The converse is also true; weak space is disconnected and open therefore to potential abuse from other parties, including unsupervised cross-oceanic navigation, terrorism and possible resource extraction (Bravo and Rees 2006).

The Arctic Ocean coastal states have responded to this sense of scrambling by articulating polar strategies and promoting investment programmes, with attendant claims that they are building more ice-breakers, commissioning more research stations, making more surveillance flights (including the use of drones) and remaining forever vigilant. Effective governance is defined and increasingly judged by those parties being able to survey, assess and administer their volumetric territory, above and beneath the permafrost, ice and sea. As with the 1940s and 1950s, when there was much interest in surveying and mapping polar territories around North America and the Euro-Asian landmass, so contemporary governments are no less interested in such things (Turchetti and Roberts 2014). Sixty years earlier, the focus was on aerial mapping and polar aviation, especially as it linked to continental defence arrangements and Cold War antagonisms (Farish 2010). As we show in chapter 3, it is now the polar seabed, especially the outer continental shelves that are being mapped, probed and evaluated with ever greater vigour in the Arctic Ocean (Brekke 2014). The inclination to map everything (as the Argentine novelist Borges might have recognized) of potential interest is due to the demands placed on coastal states to submit detailed scientific materials to the UN Commission on the Limits of the Continental Shelf, which will in due course help to determine the possible extension of sovereign rights of coastal states to additional resources on and beneath the seabed.

These attempts to quite literally fix sovereignty (or sovereign rights) on the seabed of the Arctic Ocean become all the more significant when one takes into account the spatial scale of the exercise. At stake, on the one hand lie thousands of square miles of unclaimed subterranean territory, but on the other hand, anxieties reside about the region's unstable and indeterminate geophysical properties, which are being altered by climate change. Today's ice is increasingly likely to be tomorrow's water and thus the idea of permanence becomes all the more problematic. Perhaps that explains, albeit subconsciously, the appeal of mapping the seabed (and as we discuss later, the colonization and partial settlement of the Antarctic continent). It points to something reassuringly solid in the midst of transformation at the level of the water surface. For indigenous peoples, the dynamic nature of the Arctic would come as no surprise given the sophisticated

18

understandings and knowledge of sea ice and awareness of local and regional specificities regarding wind and water currents, temperature, snow cover, ice thickness, and how they make and remake the life-worlds of those who call the Arctic home (Nuttall 1992; Bravo 2009; Hastrup 2009). This insight contributes to different kinds of understandings of Arctic regions, as we show later, as places of connections and flows (topological) rather than neatly delimited highly nationalized spaces (topographical).

This sense of connections and flows is also significant in the Antarctic even if few humans might call that place 'home'. Growing interest from countries such as Brazil, South Korea, China and Turkey has been noted with caution by those claimant states such as Argentina, Chile, UK and Australia who believe that the polar continent and surrounding ocean is under the partial purview of nationalized authority. While the Antarctic Treaty demands that all parties respect the sovereignty moratorium as noted in the provisions of Article 4, for much of the last 50 years claimant states retained an eagerness to invest in what might be thought of as a 'treaty sovereignty' (Hemmings et al. 2012). On the one hand, claimants would endorse the principles and provisions of the treaty while on the other hand seeking to consolidate their sovereign interests through investment in scientific knowledge creation (aided and abetted by scientific stations), participation in resource management schemes, overseeing and managing, where possible, other activities such as fishing and tourism, continued place naming and ensuring that their citizens were inculcated with a sense that they resided in countries blessed with Antarctic portfolios (e.g. memorials, public education and acts of commemoration).

This inculcation is all the more important when one appreciates that British children reside over 10,000 miles from the polar continent and it arguably took an invasion of the Falkland Islands in April 1982 by Argentine forces to recalibrate that intensity of connection, as British forces confronted their adversaries in the South Atlantic region. As stories about Edwardian polar explorers such as Scott and Shackleton might have faded from the popular British imagination, a 'winter war' in the far South in 1982 reinvigorated that sense of connection to one of the most distant parts of the British Empire. Much has changed in the intervening years regarding the self-governing Falklands; UK travellers can now travel to the islands by jet, stopping off at another British Overseas Territory, Ascension Island, along the way. Twenty years after the 1982 Falklands/Malvinas conflict, moreover, the UK renamed a large part of the southern portion of the Antarctic Peninsula as Queen Elizabeth Land.

19

The persistent presence of semi-claimants such as the US and Russia (they retain the right to make a territorial claim in the future) and non-claimants such as China contributes to the scrambling of the polar continent. As scholars such as Alan Hemmings (2014) have noted, territorial sovereignty in the Antarctic remains deeply disputed, and action and for that matter inaction (remaining 'silent' despite experiencing unease at the behaviour of others) contribute to this uncertainty. While the value of consensus retains considerable currency, as the maintenance of the Antarctic Treaty and its associated legal instruments depends upon it, all parties evaluate carefully their positions on contentious issues such as living resource management, outer continental shelf delimitation, the location and function of research stations, biological prospecting, public education and even the possibility of mineral exploitation in the future. Every time, for example, China establishes a new scientific station, others ponder whether this means something more sinister – is it a prelude to a territorial claim in the future? What do the Chinese mean when they say 'It is only a matter of time' with regard to the future management of the Antarctic (see Brady 2012)?

Such 'state work' in the main contributes to an enduring storyline about how the Polar Regions are being imagined as new (although we question this claim to newness) – and some media and industry commentators are saying 'last' – frontiers for oil, gas, fishing and mineral extraction. Even in the Antarctic, these frontiers are viewed as important for supplying global energy and food needs and meeting increasing global consumption demands. According to the UN Food and Agricultural Organization, around 10–15 per cent of the world's population relies on fisheries and aquaculture for their livelihoods. In 2012, over 50 per cent of the world's seafood trade originated from the global South and consumer demand across the world is expected to grow due to population growth in general, and expanding market demand in the global North in particular. The Southern Ocean represents a very attractive commercial fishing frontier, holding lucrative fish such as the Patagonian Toothfish, and it attracts multinational fishing syndicates. Illegal fishing is a major challenge for those charged with implementing living resource management and coastal states such as Australia, France, New Zealand and the UK remain active in fisheries protection but the areas in question are vast.

In the Southern Ocean, we have seen in recent time a scrambling of political consensus surrounding fisheries management. Notwithstanding the Hobart-based Commission for the Conservation of Antarctic Marine Living Resources (CCAMLR), it is now not uncommon for countries such as New Zealand to complain that other members are more interested in the 'rational use' of fish rather than 'conservation'. In contrast, those countries with an interest in fishing such as China, Spain, Ukraine and South Korea complain that 'conservation' is used to block and frustrate future harvesting. As CCAMLR is governed by consensus, the ability of parties to block proposals is not insignificant, as proponents of marine protected areas have discovered.

While Antarctica is spared at present from discussions on minerals, estimates from bodies such as the United States Geological Survey (USGS) that 25 per cent or more of the world's remaining untapped oil and gas reserves could be found in the Arctic only heighten the sense of anticipation (and the global interest in telling the story about it). Oil and gas companies talk of searching for new resources in frontier areas that are harsh and challenging, such as the High Arctic and deep-water areas. With global climate change impacting the circumpolar North in an unprecedented way, it has almost become a commonplace assumption that as sea ice melts and permafrost thaws, access to the Arctic and its resources will be easier in the coming decades than has previously been possible in the region's recent history.

The Arctic may be at the forefront of this discussion, but the Antarctic is also of intense interest as bioprospecting and fishing, and research on climate change, subglacial lakes, fisheries and geological structures attest. Media stories continue to abound about a rush for resources in the Polar Regions, last frontiers awaiting further adventure and settlement and thinly governed spaces where the worst kind of Darwinian self-interested behaviour might flourish. At the same time, the Arctic and Antarctic have been assembled in political and environmental discourses as fragile and threatened places; places that demand to be 'saved'. The purpose of this book is not to add to the repertoire of journalistic or other kinds of writing on a rush to polar resources, but to reflect somewhat critically on the nature of this discourse and what lies behind it and in front of it.

Moreover, the notion of a 'scramble', let alone 'land grab', is a reminder not only of past colonial encounters with other continental spaces, peoples and ecologies but also of their continuation (Nally 2014). A kind of extractive colonialism lives on in discussion about, and in the research that sets out to probe, remote places such as the Arctic Ocean seabed and the Antarctic continent and Southern Ocean. It is sobering to appreciate just how significant geographical knowledge has been, and continues to be, in shaping sovereign

21

power and national projects designed to bolster sovereignty and security. Arctic states such as Canada, Norway and Russia, as well as those who consider themselves to be geographically proximate such as the United Kingdom, have a long tradition of collecting, analysing and utilizing geographical information for the purpose of spatial planning, resource extraction and the development of national security strategies. In Canada, universities and learned societies were hugely important in helping to collect, codify and implement stores of knowledge about northern environments and indigenous people for the benefit of the Canadian state, whether it was in the 1950s or more recent times (Bocking 2009; Farish 2010). In Norway, the city of Tromsø is imagined as a 'gateway' to the Norwegian High North; home to a self-styled Arctic University of Norway, the Norwegian Polar Institute and recently an Arctic Council secretariat.

In the United Kingdom, with an academic and political tradition of studying 'cold places' comparatively, academic institutions and government-funded agencies such as the Scott Polar Research Institute (SPRI) at the University of Cambridge and the British Antarctic Survey (BAS, previously the Falkland Islands Dependencies Survey) are an established part of that information-gathering network, especially relevant to the British Antarctic Territory and South Atlantic islands. Nowadays, BAS also enjoys an Arctic remit, it maintains an Arctic Office in Cambridge and helps to support a scientific station in Svalbard as well as administer and manage large projects about the Arctic. Similarly, Norway, with its Norwegian Polar Institute and its Arctic/Antarctic scientific and administrative hub in Tromsø, plays a vital role in collecting, evaluating and advising successive Norwegian governments, and perhaps even more so in recent times, with a renewed emphasis on what is termed the 'High North'. Denmark also mobilized resources to study Greenland extensively in the twentieth century, with large scientific projects and field programmes in disciplines ranging from geology, glaciology, meteorology, oceanography, zoology, archaeology and physical and cultural anthropology to map, measure, codify and classify rocks, minerals, ice, water, animals and people. Recent government investment in Arctic research and the establishment of research centres and units at several Danish universities with a focus on the circumpolar North has consolidated Denmark's interest in Greenland and other northern places, especially in light of the Kingdom of Denmark's Arctic strategy. Similar claims could be made for the long-term investment in understanding cold environments by the United States and Russia/Soviet Union in both

22

Polar Regions, and Australia, New Zealand, Argentina and Chile in the Antarctic. Hobart, Christchurch, Ushuaia and Punta Arenas are actively promoted as Antarctic gateway cities.

So when we think about scrambles and scrambling, we emphasize the importance of focusing attention on the processes, actors, objects and places implicated in such activities such as the mapping and delimiting of the seabed, the discovery of objects such as submerged wrecks, the evaluation of resources, the classification and management of animals, and the relations between states and indigenous people. The technologies of calculation, such as cartography, surveying, measurement and navigation are all key to this (Elden 2013). The various chapters of this book stress the manner in which both Polar Regions are lively and contested spaces when it comes to their mapping, exploitation, evaluation and management. They are also challenging spaces where human agency does not always prevail, often defying strict spatial disciplining, and our imaginaries of the Polar Regions can fail to capture the dynamic interactions between the land, the ice, the atmosphere and the water and its role in fashioning human–environment relations (Steinberg 2013).

This sense of confusion and indeterminacy is perhaps contributing to another kind of scrambling. While there are plenty of actors eager to imagine the Arctic and Antarctic as ripe for further resource extraction and colonization, political appropriation and security, there are others warning about environmental scrambling; a view which holds that we still do not properly understand the complex feedbacks affecting polar climates and ecosystems. The fate of human and non-human objects such as sea ice and the polar bear loom large in assessments and predictions warning of a warming Arctic and an unsettled Antarctic continent experiencing warming and cooling in different geographical regions. In some geopolitical forecasting, it is not uncommon to read that the disappearance of sea ice, in the summer season for now at least, will further encourage a new scramble to exploit migrating fish stocks and non-living resources, while at the same time facilitating ever greater mobility for ships, submarines and planes to travel through, across, under and over the Arctic.

Re-scrambling the Arctic and Antarctic

In critical geopolitical scholarship, the geographical assumptions and designations routinely deployed in global politics, including those affecting the Polar Regions, remain a core concern. For those

23

scholars, geopolitics as thought and practice could not be thought of as a neutral expression of pre-given geographical facts. For those who wrote on polar geopolitics in the past, the Arctic and Antarctic were often positioned as 'exceptional spaces', exceptional in their size, location, remoteness and even their connectivity to wider global political, legal and economic networks and practices. Arguably, a new generation of scholars have re-scrambled the Polar Regions and challenged those claims to physical and geopolitical exception (Powell and Dodds 2014).

For the Antarctic, the claim to exceptionality (and even the extraordinary) is most potently manifested at the continental scale – an uninhabited land, designated solely for science and peace under the terms of an exceptional treaty, namely the 1959 Antarctic Treaty, and associated legal instruments. Another argument for exceptionality is made on the basis of geographical vastness, remoteness and harshness. For the Arctic, however, great stress is placed upon the national position – the Canadian, Russian, Norwegian and/or Danish or Greenlandic view. This is now subject to great challenge and complication as more regional (e.g. Barents, West Nordic) and indigenous visions of northern futures inform cooperative structures, land claims and demands for self-government and participation. The views expressed in subregional forums might be quite different from those held by 'southern' political elites and those views can and do rub up against one another. Larger, comparative themes are often resisted in academic scholarship, because in some ways, they are perceived to undermine the need for particular, local forms of expertise developed over scholarly lifetimes, especially in the Arctic (Powell 2010).

The controversy surrounding a photographic exhibition of the Arctic National Wildlife Refuge (ANWR) in north-east Alaska illustrates how polar geopolitics needs to be considered as complicit in a wider range of sites and spaces (Dunaway 2006). Subhankar Banerjee collated a series of images of the ANWR over a 14-month sojourn, encompassing the passing of the polar season; the long winter, the slow turn towards spring, the short intense Arctic summer, and the autumnal slide into cold and darkness. Due to be exhibited in the Smithsonian Museum in Washington, DC, his photography became the subject of geopolitical controversy when Senator Barbara Boxer, a Democrat from California, used one of his images (in this case involving a polar bear) to argue against any proposal to authorize oil drilling in ANWR.

Created in 1960, with additional lands set aside in 1980, ANWR was a contested region for several decades before this photographic exhibition was proposed. Industry has lobbied for access to drill for oil resources within the refuge, while environmentalists urge that it should remain a wilderness. Beyond the ambitions of the oil companies, supporters of 'opening' ANWR's coastal plain to drilling and development include Republican representatives in the US government, Iñupiat residents on Alaska's North Slope, and American veterans' and labour associations. Their arguments range from economic benefits to issues of freedom, identity and energy independence. With the production levels of oil fields elsewhere in the United States decreasing, the potential of ANWR as one of America's last oil frontiers has long been flagged as vital to the country's future prosperity. Proponents of development suggest that the US relies too heavily on foreign oil imports, and that this dependence creates an undesirable relationship with countries in the Middle East that harbour 'terrorists'. If ANWR was reimagined, even re-scrambled, the Alaskan Arctic in this analysis becomes a vital element in reducing reliance on imported oil from the Middle East and improving national security.

The 'open' side argues that exploration and development within ANWR will be valuable in terms of job creation and bring benefits to the local, regional and national economies. To labour associations such as the International Brotherhood of Teamsters, domestic oil creates domestic jobs, which are certainly of more value to the US and its people than workers in foreign countries earning a living from American purchases. They suggest that up to 750,000 jobs can be created in the US as a result of ANWR oil development. A pro-development website urges those sympathetic to the labour movement: 'American labor is ready to work. First you must do your job: OPEN ANWR.'[1]

In the spring of 2003, pressure was brought to bear to initiate hydrocarbon exploitation on ANWR's coastal plain in the wake of the 9/11 attacks and renewed concern about energy security. For Boxer and others opposed to any such development, the exhibition photographs provide a compelling visual testimony to its special wilderness-like qualities. The Senate eventually voted against any plans to drill in ANWR; Banerjee's photographs, and his subsequent 2003 book *Arctic National Wildlife Refuge: Seasons of Life and Land*, were identified as a critical visual and geopolitical intervention (Banerjee 2003; Dunaway 2006). This was magnified further when it emerged that the Smithsonian Museum altered their proposed exhibition of Banerjee's work, with a particular targeting of the proposed

---

[1] www.anwr.org

photographic captions. As with an earlier controversy involving the Enola Gay in the mid-1990s, an exhibition (in the heart of the capital of the United States) acted as a catalyst for a testy exchange over the way in which the Arctic and its inhabitants and ecosystems should be understood within the American national imagination. Denying any claim that the museum was the victim of overt political pressure, the revised captions attached to the Banerjee exhibition were widely regarded as 'toning down' his original descriptions in favour of concise descriptions of animals and seasonal landscapes.

The framing of Alaska and the Arctic more generally generates two opposing positions. On the one hand, the 49th state is imagined to be an enduring frontier space ('the last frontier' as Alaskan vehicle number plates remind you), filled with vast hydrocarbon potential and other resource commodities such as timber. The shale gas revolution in the Lower 48 has, however, altered the economics of Alaskan oil and gas exploitation. On the other hand, there is a long tradition within the United States of imagining Alaska as a pristine wilderness – and within the state ANWR is also represented in distinction perhaps to other parts of Alaska affected by disasters such as the *Exxon Valdez* oil spill in 1989 and the infrastructure of the Trans-Alaskan pipeline. Senator Boxer's intervention explicitly appealed to this visual and textual tradition when she deployed the image of the polar bear walking on the ice in an attempt to delay any drilling off the ANWR coastline. In our final chapter on polar demands, we posit that organizations such as Greenpeace use human celebrities such as the British actress Emma Thompson in combination with the polar bear as 'celebrity' to visualize the Arctic as a region demanding to be 'saved'.

In his thoughtful review of the controversy surrounding Banerjee, Finis Dunaway makes the important point that what got lost in the rhetorical mire was a third way of framing places such as Alaska. As he notes, 'Together, the images and texts move beyond the competing frontier visions of the ANWR – a battle over whether to preserve therapeutic wilderness or exploit economic resources – to frame the landscape instead as a place that is connected to everyday life in the rest of the United States and beyond' (Dunaway 2006: 166). In other words, a view of Arctic Alaska not as a last frontier that needs 'saving' but as a series of places and communities intimately connected to the history, geography, geology and ecology of the world itself. The coastal plain of the ANWR, for example, is a breeding ground for countless animals including herds of caribou and a staggering 180 species of migratory birds, which have travelled from as far as South America and Antarctica.

26

Indigenous peoples such as the Gwich'in, some of whom live in Arctic Village on the southern boundary of the ANWR, depend in part on hunting the highly mobile Porcupine caribou, and Iñupiat catch whales in the Beaufort Sea as part of their subsistence lifestyles. The interaction between the human and non-human world provides a reminder of how places like ANWR highlight exchange and interdependence – and that exchange, as has long been recognized, can also include other unwanted highly migratory substances such as long-range pollutants. The ANWR also illustrates frontline indigenous politics, which are entangled in land claims, movements for self-government, and visions of economic development, something we explore in more depth in chapter 5. Many residents of North Slope Iñupiat communities generally support the opening of ANWR to industry and allowing drilling on the coastal plain. They, along with residents of other Alaskan communities, look to development as a source of jobs, schools and other opportunities. As the only people living within the boundaries of the refuge, the residents of Kaktovik on the Arctic coast claim that it should be their opinion on future development that takes precedence over groups living outside. The Gwich'in, on the other hand, tend to be opposed to development. They depend heavily on those Porcupine caribou, which use the coastal plain as their calving grounds, an area the Gwich'in revere as 'The Sacred Place Where Life Begins'.[2] Opponents of ANWR development, including the Gwich'in Steering Committee, have continued to argue that it would be a fiscally irresponsible decision, since there is no way of yet knowing how much oil is available, a view shared by many environmental groups in Alaska and elsewhere in the US. They say that the financial costs incurred in exploration and development may not be recovered from oil reserves that might not be as significant as hoped; such uncertainty cannot justify damage to the land, its wildlife and its inhabitants.

Banerjee's photographs remind us how profound these struggles are. The way in which others imagine, frame and inhabit the Arctic is not straightforward, and reveals how northern residents do not necessarily agree with one another. This idea of a 'polar frontier' suggests that a coterie of environmentalists, journalists, political leaders and advocates of energy exploitation, for their different reasons, are wedded to the idea of the Arctic (in this case) as a largely untouched space, which is far from the case. The footprints of northern

[2] http://www.alaska.net/~gwichin/

27

communities, both human and animal, are there to be seen if you look and listen closely enough.

Being wedded to a particular framing, however, begins to make more sense when we consider the kind of opportunities that present themselves. For the political leader, the idea of the Arctic and/or Antarctic as a blank space has been an important element in presenting and justifying imperial and colonial strategies, which frequently lionize the achievements of those men (and it is usually men) and machines sent to conquer and occupy remote and barren spaces. For the energy executive, the frontier region represents a vision of opportunism and potential bonanza – an arena with an abundance of natural resources including oil, gas, timber and fishing awaiting exploitation and development, albeit one also encumbered with regulatory hurdles and social licensing challenges. For the environmentalists, the notion of the 'last frontier' rather than just 'frontier' signifies a hope as well as lament for the future of the Polar Regions and the planet itself. As Susan Kollin (2000) wrote about Alaska within the American geographical imagination, 'Alaska functions as an important national salvation whose existence alleviates fears about the inevitable environmental doom facing the US; like previous mystic frontiers, it promises to provide the nation with further opportunities for renewal' (Kollin 2000: 43). In January 2015, environmentalists were delighted – while Republicans and pro-developers were angered – when President Barack Obama's administration moved to declare ANWR's coastal plain, as well as other areas in the Beaufort and Chukchi seas, off-limits for offshore drilling for oil and gas (while planning to allow it in the Atlantic Seaboard and parts of the Gulf of Mexico).

The idea of the frontier and its salvation-like qualities remains an important element in understanding the contemporary framings of the North American and Russian Arctic (McCannon 1998; Laruelle 2013). In his widely celebrated account of his travels in the Arctic, Barry Lopez (1986) highlighted the way in which the idea of regions like Alaska and Canada's Yukon Territory, and increasingly others such as West Greenland, as frontiers helped to make these places more inviting to development projects. From the Trans-Alaska Pipeline in the 1970s to the exploratory drilling off the coastline of Greenland in the last decade, the Arctic emerges less as an isolated frontier region but as a transnational and neo-liberalized space connected to global flows, markets, risks and networks – the idea of either 'frontier space' or 'final reserve' is inadequate in the face of such complex interconnections. Moreover, it also serves as a cautionary coda to those who use terms such as 'frontier' without considering how that might work against indigenous communities and their understandings of outstanding land claims, subsistence lifestyles, economic sustainability and environmental stewardship (Churchill 2002; McCannon 2012). One telling illustration, perhaps, was the way in which environmental movements such as Greenpeace mobilized public opinion against seal pup harvesting in Canada and subsistence seal hunting in Greenland, and in so doing persuaded the European Union (EU) to ban the import of seal products in the early 1980s. As a consequence, it placed added importance on developing mineral extraction industries as an alternative revenue stream, one which is less likely to be distorted by trade bans because the commodities in question do not raise the same moral unease – although this is changing given environmental campaigning against Arctic resource extraction and fears of catastrophic oil spills and industrial accidents. It is still the case that considerable resentment exists in some northern communities against environmental organizations and the EU for their campaigning against indigenous resource and trading-related activities. Paradoxically, it contributed in part to the 'dash for other non-living and non-renewable resources' and led to unease about the EU being allowed to become a permanent observer to the Arctic Council.

If anything, this has intensified in the Arctic as environmental groups have demanded that the Arctic be 'saved' and protected from further mineral extraction industries, while some indigenous and northern communities complain that such demands for action reveal fundamental ignorance about the Arctic as a 'lived space', where some of its inhabitants are eager to retain their indigenous subsistence lifestyles, while others wish to improve their standards of living and quality of life and welcome resource extraction. For the residents of the Arctic, extraterritorial interest and intervention has a contradictory quality; drawing the Arctic region ever more into the everyday lives of those residing in both the southern constituencies of Arctic states and non-Arctic state residents, while 'distancing' those same people from the everyday realities of northern communities, many of which are blighted by poor health, domestic violence, modest educational attainment, inadequate housing and limited employment opportunities. Anti-whaling campaigns or initiatives to protect the Arctic from oil exploration or mineral extraction are viewed from northern communities as attempts to deny indigenous and other residents a right to development.

The subsequent chapters consider, among other things, the evolving governance of the Polar Regions with particular reference to the

Antarctic Treaty System and the Arctic Council. There is growing interest in the Polar Regions as fishing and tourist vessels ply their trade, and as even the subterranean portions of the Arctic and Antarctic enlarge (and potentially enrich) the sovereign rights of coastal states and create, by association, other areas that are part of the high seas and 'global commons'.

All of which can sit uneasily with indigenous peoples (demanding distinct recognition), environmental groups (demanding restrictions on resource extraction), scientists (demanding that extraordinary amplification be recognized), and industries (demanding access to wilderness and resources), which 'trade' in exceptionality, albeit for different reasons. But there are other kinds of demands as well: international law can be demanding, as conventions and treaties seep and infiltrate into the business of Arctic and Antarctic governance, including a forthcoming polar shipping code in 2017 alongside growing bodies of environmental law, human rights and indigenous rights legislation. Heritage and biodiversity protection also play their part in undermining a sense of geographical exceptionalism.

This poses, ultimately, a challenge for more critical geopolitical renditions of the Polar Regions and perhaps forces us to think what is at stake when claims are made regarding scrambles and scrambling. In other words, we need to think harder about how the Polar Regions are connected and networked (with all that comes with it in terms of access, demands and governance) rather than where they rigidly begin and end. We need, as our chapters address, to be willing and able to recognize and analyse the Polar Regions as complex spaces, with their own histories and futures, which are being scrambled by human and non-human intervention. Far from being remote areas of the world, they are enrolled and integrated into the daily lives of communities located well beyond any commonplace geographical definitions of the Arctic and Antarctic. Whether we choose to eat Patagonian Toothfish, listen to weather reports warning of a 'polar vortex', refuse to purchase petrol at Shell petrol stations, or study and research the Polar Regions at school and university, citizens around the world are becoming more engaged with places that not so long ago were thought by many to be at the very ends of the known world.

— 2 —

MAKING AND REMAKING THE POLAR REGIONS

The Polar Regions remain subject to different, often conflicting and contested, representations, images and ideas, which fashion the ways we think about circumpolar worlds, imagine them, live in them, venture into them, and seek to exploit or protect them. This ideational and visual repertoire furnishes cultural accounts of them as exceptional and extraordinary places to travel to and from, in which to live (and to die), to move around (and to be stuck), and to get to know (and to be confused in and about). These cultural accounts influence ideologies, objects and practices about exploration, discovery, science, settlement and development; they affect attitudes towards, and relations with, northern and southern societies; and they inform debates about the future of circumpolar lands, the seas and oceans, and peoples.

The making and remaking of the Polar Regions is of relevance to our interest in polar scrambles and scrambling. What are the drivers at play? In this chapter, we consider six drivers, mindful of overlaps and interconnections. As we reflect on them, it is apparent that there is a mixture of both human and non-human (as well as more-than-human) factors at play and that there are, as Doreen Massey (2005) noted, a whole series of power-geometries to which we need to be attentive in how they influence, shape and reproduce narratives about the Arctic and Antarctic. We do not posit a straightforward distinction between the Polar Regions as a series of exceptional local and regional spaces and the wider world elsewhere affecting the Arctic and Antarctic. As Massey cautioned, the local/global distinction is not so obvious and terms like 'globalization' need to be used with considerable care because of the ideological baggage attached to them. Yet, as we will discuss in the following chapters, the framing

31

of the Polar Regions as exceptional and extraordinary places persists, especially when it comes to thinking about frontiers, ecologies and futures.

The term 'power-geometries' was designed both to highlight inequalities between places (whether they are villages, cities, regions or continents), and as a view of places that are produced as a consequence of their positioning and relationship to other places, relationships and power-geometries. In Massey's work, London is one of the most powerful nodes in a neo-liberal power-geometry, which is made and remade through financial, labour, political and social flows that bring people, sites and objects into relationships with one another. Those relationships can and do get solidified (and even securitized) at times through the financial system, property markets, politics and the like, but they also get contested and scrambled as the Occupy movement protestors outside St Paul's Cathedral put forward alternative demands and representations of the financial crisis in 2011–12. But as Massey also noted, there are differential mobilities at stake, as political and financial elites benefit from calls to lower barriers to investment and ease restrictions on movement, while local communities complain of being 'excluded' from financial networks, property and wealth creation. Just as London's social and economic geographies more generally have been 'scrambled', we believe that there have been Arctic and Antarctic scrambles, which create, in turn, new power-geometries.

Drivers of Scrambles and Scrambling

In the Arctic and Antarctic, we identify six scrambling drivers, and trace through them some associated implications for within and beyond those spaces. Our six are: globalization, securitization, polarization, legalization, perturbation and amplification. We recognize it might be perfectly possible to identify others, but we contend that these six drivers are hugely important in the making and remaking of the Polar Regions.

Globalization

Rather than being peripheral to world events, the Arctic and Antarctic have been tied to the global economy for centuries in some cases, as well as being subject to the effects of increasing globalization (Heininen and Southcott 2010). They are global regions with global geographies and histories, which stretch widely and deeply between other regions and communities. Globalization is frequently understood to refer to the integration of the global economic system, with a particular emphasis on the intense but highly uneven flows of goods, services, money, technology, knowledge and people between and beyond national territories. In the Arctic, for example, it is estimated that something in the order of $230 billion is generated annually on the basis of intensive exploitation and export of energy sources to domestic and international markets (Heininen 2005: 91). While the equivalent amount from the Antarctic and Southern Ocean would be lower, reflecting in very large part the absence of hydrocarbon exploitation, it is a reminder that both regions are embedded within the global economic and political system, including industries such as tourism and fishing and not just the mining sector. This is not necessarily surprising given the long history of exploration and exploitation ranging from the activities of the Hudson's Bay Company and Dutch East India Company, to the Argentine, British, Norwegian, Russian, Spanish and Japanese whaling and fishing fleets harvesting in the waters around South Georgia and Antarctica itself. In South Georgia and the Antarctic Peninsula, there still exists a network of abandoned communities where the artefacts of the whaling industry remain, such as whale oil storage tanks, landing stations and community infrastructure including churches. From these whaling stations, whale oil was transported to European markets and provided, at one stage, a vital source of heating to faraway cities.

The Arctic has a longer history of global convergence and economic integration than the Antarctic. The fur trade is documented back to the ninth century in the Eurasian North, and first brought northern peoples into contact with traders from regions to the south (fur was soon coveted by people living in Egypt and the Chinese were demanding furs in the sixteenth century – there is much still to write on the history of Chinese traders in Siberia, for example). Also in the ninth century and later in the tenth, Norse voyagers sailed west from Norway and the British Isles across the North Atlantic to settle in Iceland, and then to south and south-west Greenland. Soon the northern North Atlantic, from Greenland across to northern Norway, was linked through extensive and wide-ranging trade routes throughout Scandinavia, Europe, Russia and Greece, and with Arab traders from further east, but also from Spain and North Africa (Seaver 1996).

The North Atlantic became a busy seaway. Vessels from Norway brought timber and other necessities to Iceland, while Iceland supplied the Greenland settlements and farms with meal and corn. Ships sailed across the northern seas from Iceland and Greenland to Europe with seal oil, woollens, falcons, polar bear furs and walrus ivory, all of which entered even more distant markets. Goods and objects from the North would find their way along medieval trade routes, moving along slowly on the cold, grey seas and on rivers and roads, exchanging hands in various markets filled with an eclectic mix of northern and Asian commodities – gold, silks, spices, wine, ermine and sable furs, birchbark, and slaves, baleen and the occasional whale or polar bear skull. Kirsten Seaver's account of Norse Greenland and its place in the wider North Atlantic and beyond gives us a glimpse of the extensive nature of trade between Greenland, Europe and further afield, and of the demand for northern goods. Norse Greenlanders would go on journeys to the far north of the island on walrus hunting expeditions, and walrus ivory soon entered the trade economy of western Europe, but it also was the currency in which the Greenlandic churches paid their tithes to Rome. Some trade items from Greenland were of such good quality that they were prized as luxury items, and walrus tusks and Greenland falcons – white gyrfalcons – were particularly sought after. 'It is probably safe to assume', writes Seaver, 'that neither ivory nor gyrfalcons were *ever* traded cheek-by-jowl with codfish and sheepskins in the Bergen market' (Seaver 1996: 82). Flanders, too, was prominent in medieval European trade and had important connections to Greenland and the Norse walrus hunters long before Dutch and other European whalers turned their attention to Greenland's icy waters.

Other northern trade routes developed, such as that of the Pomors in north-west Russia. Archangelsk, on the White Sea coast of northern Russia, became a centre of the walrus ivory trade, and in the 1700s its craftsmen (many of whom were sent for several years of apprenticeship in Moscow and St. Petersburg) produced sophisticated carved ivory objects, including caskets, toilet boxes and combs – all of which were in high demand in Europe. Much of early Copenhagen was built from profits of the Danish trade with Iceland, while by the 1780s the Royal Greenland Trade Company presided over a social class system in which an upper social stratum of indigenous Greenland Inuit played an active role in whaling, and in the trade in seal skins and blubber. As a result of such mercantile encounters, a new society emerged in Greenland, based largely on the production and trade of marine resources that found expression in many cultural forms and nurtured and celebrated a national identity by the nineteenth century. Danish and Norwegian commercial activity along the Gold Coast of West Africa from the mid-1600s to the beginning of the nineteenth century relied heavily on salted and dried fish, and North Atlantic fishing communities were involved – unwittingly perhaps – in the production of dried fish that became an important item in the slave trade, both as a food item in West Africa, but also on the slave ships that made their way across the Atlantic to the Caribbean. Thus, we can begin to trace transnational connections between small northern coastal communities and the growth of plantations in the Danish West Indies. In the North Pacific, until Europeans disrupted Native economic activities, Alaska was involved in extensive trading relationships with Siberia and diverse cultures and economies were linked in a network that stretched a vast geographical distance, across Siberia and south to Korea, China and Japan. In Alaska, archaeologists have found ornamental objects from Asia, trade beads and tea, and evidence of Chinese influence in art, stretching back some 2,000 years (Fitzhugh and Chaussonnet 1994; Bockstoce 2009). The point is that these spaces have their own histories and geographies of connection and interaction.

At the opposite ends of the earth, Cook's descriptions of seas teeming with marine mammals led to a rush to the Southern Ocean by sealers, whalers and explorers. The South Shetland Islands were discovered in February 1819 by sealer William Smith and a sealing boom followed. The British Admiralty engaged Smith to survey the islands under the command of Edward Bransfield and they both saw and charted part of the Antarctic Peninsula in January 1820. Further mappings were made by other expeditions and, on 7 February 1821, the crew of the America sealing vessel *Cecilia*, under the command of Captain John Davis, made the first recorded landing on the Antarctic continent at Hughes Bay. By the time of this voyage, British and American sealing ships were already active in the South Shetlands. Their activities intensified and the fur seal population had seriously declined by 1830. Antarctica was very much caught up in the international rivalry, especially between the British and Americans, in commercial sealing and whaling enterprises. European geographical societies, established between 1821 and 1830, played a significant role in promoting polar exploration, but many of the voyages that now went south to Antarctica and charted portions of the Antarctic continent were led by people connected to whaling interests. Apart from a few exploratory voyages, such as that of James Clark Ross, who discovered the Ross Sea and established that the South Magnetic Pole was inland in 1841, Antarctica was almost the sole preserve of sealers and whalers until the final decades of the nineteenth century. Towards the end of the nineteenth century, modern whaling firms had improved the commercial potential of the Antarctic whaling industry and, in the early twentieth century, whaling became the chief economic reason to explore Antarctica and resulted in the first claims to territorial sovereignty by the British in 1908 and 1917, although this was overshadowed somewhat in the first years of the century by various expeditions – the last of the heroic age of exploration – aiming to be the first to reach the South Pole and make other polar crossings.

British and Norwegian whaling in the Antarctic was instrumental in shaping the future geopolitics of the region (Roberts 2011). Under British administration, Norwegian whalers established communities in places such as South Georgia and the Antarctic Peninsula. European bodies, objects and technologies moved southwards, and migrated around those polar waters in their search for whales. Whale products including oil were transported to European markets, and British colonial control became dependent on whaling revenues for the continued occupation of what was described as the Falkland Islands Dependencies. For the first half of the twentieth century, whaling funded British mapping and surveying of terrestrial and maritime polar environments. The end of commercial whaling in the 1960s combined with anti-whaling campaigning transformed British polar geopolitics, and the Antarctic's connections with the wider world (Dodds 2002). One noticeable feature of that shift was the abandonment by Norway of their whaling settlements in South Georgia, which resulted in British anxieties that unless they acted to re-occupy the island, then Argentina, a counter-claimant, might act. In March–April 1982, Argentine forces did indeed invade and occupy both the Falkland Islands and South Georgia. Where the whaling industry once provided a secure presence in the form of invited Norwegian partners, the British were forced to rely on the military and later scientists (in the form of a new station at King Edward Point) to deny different kinds of connections with Argentina from taking root.

The key point to be noted here is that the histories and geographies of the Arctic and Antarctic are both intensely local but also global in terms of flows and networks. People and objects flowed backwards and forwards. Things got traded. People moved and settled – and, in places right across the northern circumpolar world, were also colonized, moved and settled by the state. Contaminants floated and settled in and then under the ice, as we note later. Markets made a difference to communities and the rise and fall of market demand for products derived from whales and seals generated their own particular geographies of money and legal geographies as fur seal-rich regions such as the Bering Sea became the object of legal dispute, arbitration and finally convention in 1911. Changes in fashion and environmentalist discourses in the closing decades of the twentieth century also had a profound impact on the histories of indigenous and northern communities, as seal product bans led to a collapse in certain markets.

Our recognition of globalization as a driver, then, is not predicated on novelty. The Arctic and Antarctic, especially in the last two centuries, have been ever more closely tied to flows of people, money, ideas and influence. Both the human and non-human are caught up in these flows and their interactions with territories. The objects, agents and processes that help constitute those territories and flows have human and non-human aspects such as the shifting currents and flows of ocean and ice, the seasonal distribution of animals, and the short- and long-term fluctuations in commodities markets. But that form of globalization is deeply contested, the mobility of things, people and the like produces a series of entanglements, some of which might be welcomed by indigenous and non-indigenous stakeholders and some of which is not. In the Arctic, this is perhaps most vividly demonstrated by the assemblage of debates, negotiations and protests by northern communities as they, and their national governments, negotiate with mining companies about mineral prospecting, development and exploitation. For some, 'opening up' the Arctic to global energy and mining corporations offers opportunities for local employment and socio-economic improvement, while others contend that new dependencies will materialize through the relationship with foreign-based operators and the capriciousness of global resource markets themselves. These are not unique fears, as other communities around the world have struggled with reconciling such things. As experiences in Alaska, northern Canada, Greenland, Norway, Sweden, Finland and the Russian North attest, it can all sit rather awkwardly with localized ideas about community cohesion, social licensing and the protection of traditional and subsistence lifestyles.

While globalization is complex and multifaceted, attention is drawn to the contemporary impact of various flows and exchanges within and beyond the Polar Regions. These include the exploitation and utilization of natural resources, including fish, oil, natural gas and strategic minerals such as iron ore, zinc and uranium. Again, while acknowledging the historical antecedents of mining in the Arctic, the last 50 years have witnessed a growing entanglement of mineral resources, state development, geopolitics and global markets, which have combined to make the exploitation of the North Slope of Alaska, the Barents Sea and northern Russia financially and geopolitically possible. The growth of tourism has also seen thousands of European, North American and East Asian visitors travel to the Antarctic and places like Alaska and Svalbard, generating new revenue streams and economic opportunities for local communities and foreign investors alike. Where once visitors came to the Arctic and Antarctic to hunt seals and whales, now they come to pay to watch these creatures inhabit their seasonal homes.

Securitization

Security, as Simon Dalby reminded his readers, is an essentially contested concept (Dalby 2009). Whilst traditionally, practices of security revolved around the military, since the end of the Cold War security was increasingly being used in a variety of ways, including embracing economic, environmental and human security as opposed to the traditional military-orientated focus. As such, the framing of 'security issues' in the first place was always worth a second glance. While the act of securitizing may be a way of seeking and receiving additional resources, it was notable how a wider range of issues and problems were being understood as security issues in the 1990s onwards. These included environmental change, disease, community vulnerability and humanitarian intervention.

Securitization is a term used by critical security studies scholars to refer to the manner in which objects (whether it involves countries, regions, objects, infrastructures, resources and/or populations) are secured. In order to understand the process, as opposed to the end result, of securitization, the manner in which invocations of danger, threat and risk are used to appeal to the need for political and financial resources deemed necessary for the task in hand is significant. As a consequence, such securitizing interventions bring to the fore power-knowledge relations, as claims are made regarding who knows best. This has taken added resonance, in recent years, when one factors in how the logic of anticipation and pre-emption in the face of a possible looming attack or disaster has been appealed to. States and governments increasingly solicit the need to act early and often if they think there is a need to block, prevent and dismantle those who might threat their citizens, their ecosystems and their infrastructures.

This critical literature on security suggests that claims to security are never politically innocent. One powerful example is provided by the ongoing controversy involving Australia and Japan over whaling in the Southern Ocean. While the Japanese continue to engage in whaling, under the auspices of the JAPRA II research programme, successive Australian governments have challenged these activities as injurious to Australia's environmental security as epitomized by the creation of the Australian Whale Sanctuary, including the waters off the Australian Antarctic Territory. In the last few years, a range of Australia-based actors have either mounted legal challenges or engaged in direct action, such as the Sea Shepherd Society, in an attempt to halt the harvesting of whales. While Australia mobilizes environmental security discourses to position itself as a responsible steward of the Southern Ocean, it is also worth noting how these interventions have the direct effect of cementing and championing Australia's status as claimant state within Antarctica, including the Southern Ocean. The International Court of Justice eventually ruled on the issue of Japanese-sponsored whaling and found in their judgement of May 2013 that it did not have an adequate 'scientific' basis (as demanded by the International Whaling Commission). It was thus considered a commercial enterprise, contrary to the IWC-regulated Southern Ocean Whale Sanctuary (Scott 2014).

What was interesting about the whaling controversies affecting Japan and Australia was how images of hunted whales in particular played a crucial role in bringing the remote yet inhabited Southern Ocean into the everyday lives of Australians and others. One might have been forgiven for not realizing that Australia once possessed a whaling industry and was a committed supporter of its whaling interests in the Southern Ocean. Since the 1960s, however, as anti-whaling sentiment gathered public support, successive Australian governments have tended to be very critical of any commercial hunting of whales. As a consequence, the whale became an object of, and for, stewardship rather than a renewable resource, which Australia had to securitize in the name of defending its territorial and resource interests.

Australia is not alone in securitizing polar-based resources. In the High North, recent Russian policies and strategies illustrate well this securitizing imperative. Under President Putin, Russia has identified the Arctic as a strategic priority as demonstrated in major policy statements such as the 2008 'Foundations of the Russian Federation's state policy in the Arctic until 2020 and beyond'. The main objectives revolved around the notion that the Russian Arctic was a strategic resource base and that securing the region was a national strategic priority. The future of the Russian Federation was directly tied to the energy sector and transportation/infrastructure development, which would allow the Arctic to be managed and monitored by the

Russian armed forces. In 2013, Russia adopted the 'Russian Strategy of the Development of the Arctic Zone and the Provision of National Security until 2020'. As with the 2008 statement, emphasis was given to securing sovereign rights in the maritime Arctic, funding and implementing major oil and gas projects and obtaining investment for the necessary transport and infrastructure needs relating to making the Arctic a secure resource space. Particular places such as the Kara and Barents seas were identified as crucial to realizing this vision, as they are thought to contain vast resources of oil and gas. Onshore, the Yamal Peninsula is widely regarded as the most significant in the Russian Arctic, with estimated gas reserves of around 11 billion cubic meters and oil resources of 4 million tonnes with reserves amounting to around 2 million tonnes.

As part of that securitizing move, the Russian authorities have assembled a series of measures designed to secure those resources, both actual and potential. One strategy is exercising control of how foreign companies do business with the state-owned companies of Gazprom and Rosneft, which themselves are in control of continental shelf projects. Another is more explicitly military-related, as Russian armed forces increase their presence in the region, and establish a new military command for the North. New bases are to be established at offshore islands such as the New Siberian Islands, and a primary goal will be to ensure that the Northern Sea Route (NSR) is protected, as its usage by international operators expands. It is worth bearing in mind that the NSR was largely closed to international maritime traffic before the early 1990s. Securitizing the Russian Arctic is also costly; new investment is needed in search and rescue, port infrastructure, refuelling, icebreaker support and oil spill disaster response.

But there are also limits to securitization. Russia needs foreign direct investment and expertise in order to explore, exploit and transport oil and gas in the maritime Arctic. The development of the continental shelf will require at least $500 billion and take decades before oil and gas is exploited in large-scale and profitable quantities. Russian operators need foreign partners, whether Western or Asian, and in 2011 Exxon-Mobil agreed to partner with Rosneft to pursue opportunities in the Kara Sea. Norwegian, Chinese, Vietnamese, French and Italian companies are also involved in resource investment and revenue sharing partnerships. Global energy markets and oil and gas prices will also play a key role in shaping future resource development, regardless of how far a state such as Russia seeks to defend and protect its offshore resource rights in the Arctic. On the one hand, the shale gas revolution in North America has transformed energy markets, making pipeline ventures such as the Mackenzie Gas Project in Canada's Northwest Territories uncertain, and on the other hand, major energy companies such as Shell have postponed their exploratory work in other parts of the Arctic (although the prospects of the Canol shale oil formation in the Sahtu region of the Northwest Territories have hinted at new kinds of resource ventures in Canada's North). As Russia and its international partners have found in the midst of the Ukrainian crisis, the effect of sanctions and worsening geopolitical relations can make all of this harder and indeed more precarious.

Russia and Australia are not alone in trying to securitize resources, living and non-living. Underlying the fate of whales in the Southern Ocean and oil and gas deposits off the Arctic continental shelf is a desire by proximate states to defend and monitor 'their' waters, continental shelves, and even air space, as Arctic states such as Canada and Finland watch with concern as Russian military jets approach their territorial limits. While the sovereignty regime of the Southern Ocean is rather different to the Russian continental shelf, the impulse to securitize remains a powerful one.

Legalization

One of the most important legal documents pertaining to the Polar Regions is the United Nations Convention on the Law of the Sea (UNCLOS), which entered into force in 1994. Negotiated over the 1970s and early 1980s, it is often referred to as providing a blueprint for the governance of the world's oceans and seas. The zoning of ocean space, as Philip Steinberg (2001, 2010) has noted, is one of the most important legacies of UNCLOS, as the negotiators determined the rights and responsibilities of coastal and non-coastal states. From determining a territorial sea to establishing an exclusive economic zone (EEZ) to delimiting the outer continental shelf, sovereign state jurisdiction sits at times uneasily with the ocean either as a free space of flows and/or a space of dynamic physical processes (figure 2.1).

The Russian flag-planting on the bottom of the central Arctic Ocean in the summer of 2007 provides a powerful illustration of how legalizing drivers work to make and remake the Arctic as a governed space. Once photographs of that Russian flag attached to a titanium pole began to circulate, so headlines warning of a new scramble for the Arctic proliferated (to which we will return in chapter 4). Regardless of the fact that this privately funded expedition was part of a wider oceanographic expedition undertaking the kind of scientific

41

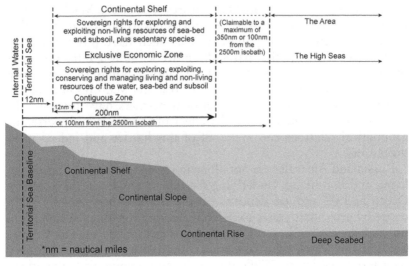

Figure 2.1 The United Nations Law of the Sea Convention (modified from www.dfo-mpo.gc.ca/oceans/canadasoceans-oceansducanada/images/ maritime-eng.jpg)

research demanded of coastal states by UNCLOS, the visual impact of a national flag being planted at the North Pole was immense.

There is a long-standing visual archive of explorers, mountaineers and even astronauts, planting national flags on the polar continent, the North and South Poles, mountain peaks such as Everest and the moon. While the Apollo 11 expedition was careful not to claim the moon as American territory in July 1969, the symbolism of planting a national flag was not lost in an era when exploration, science and geopolitics were Cold War bedfellows. In 2007, while international lawyers were swift to dismiss the Russian flag-planting gesture as somewhat baseless, it seemed to touch a proverbial nerve at least with Canadian political figures. Former Foreign Minister Peter Mackay informed an audience that, 'The question of sovereignty of the Arctic is not a question. It's clear. It's our country. It's our property. It's our water.' Both Canada and Russia have had a long tradition of imagining themselves as Arctic states with a provenance that stretches, sector-like, all the way to the North Pole. So what un-nerved Canadian political figures, therefore, was the perception that the presence of another country's national flag inferred that the North Pole and the central Arctic Ocean was 'up for grabs'. And it is worth noting

42

that, at the time, the issue of the ownership of Hans Island in Nares Strait between Ellesmere Island and north-west Greenland, something contested between Canada and Denmark/Greenland, was also preoccupying Canadian federal ministers. The flag-planting incident conjured up a vision of a subterranean place full of possibilities for further discovery, exploration and even exploitation on and below the surface of the seabed.

The Canadians probably thought they were right to be concerned, as a flag planted in the Arctic has a certain ring to it. Certain things have tended to follow one another such as planting a flag, naming a place, followed by mapping, claiming and exploiting a territory. For at least 400 years, European commercial, academic and political sponsors have sought to discover new trade routes and resource opportunities as well as fund and support expeditions designed to further exploration, discovery and exploitation. The Arctic is not alone in this association. A glance at a map of Antarctica, for example, would reveal a 'Coates Land', which was a legacy of commercial and philanthropic sponsorship of the Scottish National Antarctic Expedition (1902–4) by the wealthy Coates family in Edinburgh. Commercial sponsorship went hand in hand with exploration and science and the 2007 expedition, which led to the Russian flag being deposited on the seabed of the Arctic Ocean, had its own private sponsorship.

Following the flag incident, and despite many denials about its relevance and importance in international law, it did contribute to a more prominent reflection amongst the five Arctic Ocean coastal states of Canada, Denmark/Greenland, Norway, Russia and the United States about the current legal status of the Arctic Ocean. After some negotiation led by the Danish government, a declaration in May 2008 – after a meeting held in Ilulissat on Greenland's west coast – declared their public commitment to the 'Law of the Sea' as their preferred legal mechanism to settle any outstanding conflicts or uncertainties relating to the sovereign rights of the Arctic continental shelf. Unsettled by repeated media framings that there was a 'scramble for the Arctic', there was a deliberate campaign to emphasize Arctic coastal state cooperation. The declaration was a 'calming measure'. The agreed text sought to dispel the scrambling rhetoric:

By virtue of their sovereignty, sovereign rights and jurisdiction in large areas of the Arctic Ocean the five coastal states are in a unique position to address these possibilities and challenges. In this regard, we recall that an extensive international legal framework applies to the Arctic Ocean as discussed between our representatives at the meeting in Oslo on 15 and 16 October 2007 at the level of senior officials. Notably, the law of the sea provides for important rights and obligations concerning the delineation of the outer limits of the continental shelf, the protection of the marine environment, including ice-covered areas, freedom of navigation, marine scientific research, and other uses of the sea. We remain committed to this legal framework and to the orderly settlement of any possible overlapping claims (Kingdom of Denmark 2008).

While the declaration was intended to remind the wider international community about the role of the Law of the Sea and their collective rights as littoral states to govern the Arctic Ocean, it was also designed as not just a calming measure but perhaps something akin to a submarine counter-measure. The words and deeds of the Law of the Sea were not passive accomplices in all of this but something to be 'lobbed' at those who wished to assert their claims in support of 'the scramble for the Arctic'.

What made this necessary in the eyes of the Arctic 5 was that others, including members of the European Parliament, were proposing different kinds of legal frameworks, some of which were informed by the experiences of the other pole. The 1959 Antarctic Treaty, with its focus on sovereignty suspension and international cooperation was being championed as a model that should be emulated by Arctic states and others who had a vested interest in this maritime region. The five Arctic Ocean coastal states explicitly rejected this analogy, and as the declaration noted, 'This framework provides a solid foundation for responsible management by the five coastal States and other users of this Ocean through national implementation and application of relevant provisions. We therefore see no need to develop a new comprehensive international legal regime to govern the Arctic Ocean. We will keep abreast of the developments in the Arctic Ocean and continue to implement appropriate measures.' The Law of the Sea was the counter-measure designed to destroy the words and deeds associated with the Antarctic Treaty.

While the Arctic 5 acknowledged that there was more to be done in terms of environmental stewardship, fishing and freedom of navigation of the Arctic Ocean, the intent of the declaration was to reject the notion that this region needed a new comprehensive Arctic Treaty. While Article 234 of UNCLOS acknowledges the rights of coastal states to impose more stringent measures to ensure higher standards of shipping safety and environmental protection, other organizations such as the International Maritime Organization have undertaken to provide new guidance and binding rules for ships working in ice-covered waters. The Polar Code, which should enter into force in January 2017, will coexist alongside coastal state intervention by Canada and Russia to regulate shipping through the Northwest Passage and Northern Sea Route, respectively.

Coastal states such as Canada and Russia have used Article 234 to intensify their sovereign authority over ice-filled areas via legislative, constabulary and surveillance-based initiatives and interventions. Some countries including the United States had misgivings about Article 234 at the time of the UNCLOS negotiations precisely because there was a concern that coastal states would use 'environmental protection' as a pretext to impinge upon freedom of navigation. For example, Canada passed the Arctic Waters Pollution Prevention Act (extended in 1985 to cover 200 nautical miles from the Canadian baseline), which was designed to impose greater restrictions on those seeking to navigate through the Northwest Passage. It has also insisted that foreign vessels participate in NORDREG, a database of all vessels entering Arctic waters in the proximity of Canada. The Canadian government is also investing in a Northern Watch project designed to help the country literally 'listen' for users of Canadian 'internal waters' via a network of underwater sonar devices. In the context of the NSR, Russia insists upon 'controlling all maritime traffic within 200 nautical miles of its Arctic coastline' (Flake 2013: 45) and stipulates that all vessels transiting the NSR have to obtain prior permission from Moscow and follow a series of regulations including escorting. Between 2012 and 2013, the Putin government issued a series of directives, which reinforced these requirements and established a new NSR administrative body to oversee this regulatory structure. Both countries are committed to either monitoring and, where possible, administering the movement of commercial shipping. With further predictions of an ice-free Arctic in the coming decades, the relevance and applicability of Article 234 is likely to come under closer scrutiny. Navigation rights in the Arctic Ocean might well be a far more divisive issue in the future as opposed to resource rights, which grabbed the global headlines in 2007–8.

As Greenpeace discovered in September 2013, Arctic coastal states such as Russia pursue their security projects with force if necessary. The Greenpeace ship *Arctic Sunrise* attempted to travel close to a Russian-owned oil platform in the Pechora Sea. The so-called 'Arctic 30', after their arrest on charges of 'piracy' by Russian security forces, were participating in a Greenpeace campaign dedicated to promoting the idea that there should be an oil and gas moratorium in the Arctic Ocean. Linking oil and gas exploitation in the Arctic to global climate change, Greenpeace had previously sought to board oil platforms off the coastlines of Canada, Greenland and Norway. Despite worldwide protest against the campaigners' detention and imprisonment, the Russian authorities eventually changed the charges from 'piracy' to 'hooliganism', and they were eventually freed some three months later. President Putin pardoned the Arctic 30 just in advance of the 2014 Winter Olympics in Sochi.

The Arctic 30, composed of men and women from some 18 countries, serve as a reminder how domestic legislation was used to justify Russian securitization even if there were plenty of international lawyers and institutions such as the International Tribunal for the Law of the Sea, which posited that Russia failed to follow proper maritime legal procedures. Legalization has not only been something impacting northern seas and oceans, but has arguably been hugely significant in shaping indigenous and non-indigenous relations in the Arctic.

Whether we consider the provisions of the 2007 UN Declaration on the Rights of Indigenous Peoples (UNDRIP) – which note that indigenous peoples enjoy resource rights over their lands – and/or land claim and devolution agreements in northern Canada, the role of national and international law has been hugely impactful. From the 1970s onwards, a series of land claim agreements transformed the legal and political geographies of North America. In 1993, for example, the Nunavut Land Claims Agreement meant that Canadian Inuit became a huge landowner and with that enjoyed access to mineral and other resources. In 2010, Greenlanders were legally entitled to the subsurface mineral rights and in so doing have literally 'fuelled' fundamentally different debates about greater autonomy and possibly independence rather than dependence and neo-colonialism.

Our point is that the changing legal geographies of the Polar Regions are having a major impact on how the Arctic and Antarctic become embedded within national and international legal regimes, and how that affects both the imaginative and material qualities of these places. One of the most powerful counter-points to the idea that there are a series of 'scrambles' affecting the Arctic and Antarctic is a recognition that legal systems are creating an ever more complex mosaic of governed spaces, which if anything are becoming 'thicker' legally speaking rather than 'thinner'. While we might identify 'gaps' in marine governance, the reality is that the Arctic and Antarctic are not thinly governed, and not lightly touched by legalization. There is every reason to believe that legalization will intensify spatially and functionally. As the central Arctic Ocean becomes more accessible, international regulations relating to the high seas and seabed will take practical effect. At the same time, legal instruments pertaining to biodiversity and wilderness in areas beyond national jurisdiction become ever more significant.

Polarization

Polarization is a neologism used here to highlight, on the one hand, the growing activism of indigenous peoples and circumpolar cooperation in the Arctic involving subnational governments and regional organizations and, on the other hand, the expanding interest of extraterritorial actors such as the European Union. While the Arctic still bears its Cold War legacies, including the militarization associated with the two superpowers, the United States and former Soviet Union, it is much changed from the early to mid-1980s. Decolonization, and more recently devolution, coupled with the reduction of Cold War tensions, facilitated demands for greater autonomy, self-determination and regionalism. The creation of the Canadian territory of Nunavut, the establishment of Home Rule (and more recently, Self-Rule) in Greenland, and ongoing land claim agreements and negotiations (and disagreements over implementation), especially in the North American and Nordic Arctic, have contributed to a new impetus for a distinctly 'northern governance'.

But it has also revealed schisms. The application of the EU to be an observer to the Arctic Council is a case in point. Rejected in 2009, the EU had been developing an Arctic policy since 2008, after earlier involvement in what was termed the Northern Dimension in 1999 following the admittance of Finland and Sweden and border interaction with Russia. Although Greenland left the EU's predecessor the European Community in 1985, the EU as a regional organization has two Arctic member states (Finland and Sweden) and Greenland remains part of the Overseas Countries and Territories associated with member states. As part of that desire to develop an Arctic policy, the European Parliament alienated Arctic states such as Canada and Russia by calling for a new treaty in 2008–9 designed to enhance the protection of the Arctic, initially for the unclaimed and unpopulated area of the central Arctic Ocean. Disturbed by reports that some Europeans might have thought the maritime Arctic to be a thinly governed space, the five Arctic Ocean coastal states were swift to reiterate their primary role as stewards of the Arctic Ocean with sovereign rights over various portions of that oceanic body. Compounding that sense of unease was a sense of irritation that EU regulations on trade in seal products interfered with indigenous and northern communities' lifestyles and livelihoods. In particular, with the support of its indigenous peoples, Canada was vocal in its opposition to the EU becoming a permanent observer to the Arctic Council. The fate of the EU as observer to the Arctic Council remains unclear as we noted earlier. At the Arctic Council's 2013 ministerial meeting, the EU application was not considered because of the time taken to approve new state observers including China, South Korea and Singapore. The decision was deferred but it remains polarizing for countries such as Canada and Russia, who were reluctant to approve, in contrast to the EU's Arctic champion, Finland. What this story reveals is that the eight Arctic states have their own interests and values regarding the Arctic and that the EU is a complex candidate because of its diverse stakeholders and capacity to intrude into the Arctic region through knowledge creation, legal regimes, trade regulation, resource extraction and political presence in a multiplicity of arenas. EU sanctions against Russia following the annexation of Crimea and continued unrest in eastern Ukraine might well end up ensuring that the EU's application will not be approved at the next ministerial meetings in 2015 let alone in 2017.

But the EU is not the only actor to be deeply polarizing as Greenpeace's activities in the Arctic demonstrate. While Greenpeace, in alliance with others such as the Antarctic and Southern Ocean Coalition, was active in the Antarctic in the 1980s when debates raged over the desirability of a potential mining regime (the CRAMRA negotiations 1982–8), it was their campaign against Arctic seal hunting that really brought them into conflict with indigenous communities. In the High North, the name Greenpeace still polarizes opinion. The most notorious example involves the environmental group's attack on the moral legitimacy of seal hunting in the 1970s and 1980s (Wenzel 1991; Lynge 1992). Inuit had a long history of selling seal pelts to foreign traders, and seal meat and skin are essential elements to subsistence living. Others including non-Inuit communities, however, were also involved in seal hunting, especially the photogenic harp seals in northern Canada, Russia and Norway.

By the early 1970s, the United States instituted a trade ban and this was followed by the Europeans in the early 1980s and finally by Canada in 1987. The result was devastating, even if some exemptions were offered to Inuit hunting. The value of seal pelts fell, and adult seal products plummeted in value alongside the highly charged visuals of harp seal hunting. The income of Inuit communities dropped in Canada and Greenland, and it is widely believed that the collapse of seal hunting acted as a catalyst for social dislocation, economic hardship and worsening mental health problems for men, who traditionally dominated the seal hunt. Hunting and trapping more generally was also targeted by environmental campaigners, resulting in further resentment of how extraterritorial NGOs could intrude and profoundly disturb (see below) indigenous lifestyles, including ones involving mechanized transport and rifles.

While Greenpeace has apologized for its historical involvement in the seal trade ban, there is still considerable resentment at how the 'Save the Arctic' campaign is premised, so many northern communities believe, on a lack of consultation with Inuit and other indigenous communities. One thing that has changed since the 1970s is that Inuit are substantial landowners, are engaged in their own corporate enterprises, and have vested interests in fishing and mining, even if there are still divisions over whether mineral exploitation should intensify further. Fundamentally, all Inuit agree that they should be the primary decision makers about how their lands get developed and exploited rather than being told by others. When Greenpeace initiated protests against Greenlandic oil development, they in effect challenged a decision made by the Greenlandic government to license oil exploration and exploitation. Inuit representatives wondered, as a consequence, why 'saving the Arctic' gave Greenpeace the right to interfere with a local political decision. In other areas of the Arctic such as Baffin Island in Canada, however, some Inuit groups have cooperated with Greenpeace in order to protest about seismic testing, perhaps recognizing that environmental groups can help publicize local grievances and misgivings.

Our point is that economic development in the Arctic is polarized. Not all Inuit and other northern indigenous communities are fully supportive of resource exploitation. While some indigenous communities despise Greenpeace and other environmental groups, others can create strategic alliances in order to promote their own agendas and developmental goals. States, corporations and regional organizations are also caught up in this polarizing. Greenpeace Canada, for example, accused the Canadian government of pursuing a pro-oil agenda during its chairmanship of the Arctic Council (2013–15), whilst Greenpeace pursued the toy company Lego over their collaboration with Shell, after being deeply critical of their oil-related activities in Alaska. In the Arctic, environmental groups more generally often act as a lightning rod for these conflicts and disputes over how the Arctic is imagined (e.g. wilderness or industrial/resource base) and managed (e.g. as the exclusive sovereign concern for Arctic states and indigenous peoples).

Perturbation

Perturbation, used here in the sense of highlighting disturbance, is a critical driver. Put bluntly, when situated within debates concerning catastrophic climate change, the Polar Regions are routinely represented as either 'canaries in the coal mine' and/or the geographical manifestations of significant 'tipping points' and the crossing of 'thresholds'. In other words they are represented as 'perturbations' that stimulate substantial shifts in the warming process, made worse by positive feedback mechanisms, with dire implications for ice cap stability in Greenland and Antarctica in particular. For public commentators such as Clive Hamilton, and his *Requiem for the Species* is sobering reading, the continued upward trend involving greenhouse gas emissions is making the possibility of stopping, let alone slowing down, climate change truly fanciful (Hamilton 2010). The world continues to carbonize and industrializing countries such as China are expected to be responsible for over 30 per cent of world greenhouse gas emissions by 2030.

Both the Arctic and Antarctic play crucial roles in global climate dynamics, and both regions are experiencing rapid physical changes in response to global warming. Scientists have long understood the importance of studying the polar regions because of the ways they influence the earth's weather systems – for example, the sea ice in both the Arctic and Antarctic is a major element in the global climate system, while the Southern Ocean plays a significant role in processes of biogeochemical cycling and the exchange of gases between the ocean and the atmosphere. Meteorological stations run by several national Antarctic scientific programmes show the Antarctic Peninsula has experienced strong and significant warming over the last 50 years, while Antarctic sea ice extent has also been affected, although satellite data since 1978 show regional trends rather than a ubiquitous trend in sea ice duration – for example, sea ice duration has increased in the Ross Sea, but decreased in the Bellingshausen Sea (Chapin et al. 2005; Lemke et al. 2007). Yet while observed changes in the mass balance of the major Antarctic ice sheets and the breaking up of Antarctic ice shelves are of significant concern, the attention of a considerable number of polar science initiatives is currently focused on understanding what is happening in the high latitudes of the circumpolar North.

The Arctic Ocean is routinely cited as a place most likely to cause alarm and indeed shock amongst the climate change and policy-making communities. Observations of the lowest recorded extent of sea ice in September 2007 and September 2012 led some scientists to

50

argue that the previous distribution cannot be retrieved if scenarios of ice-free summers are realized in the very immediate future (Wadhams 2012). Sea ice scientists can also disagree about the scale and pace of change. While a majority of sea ice experts do not believe the Arctic Ocean will be free of ice until the 2030s or 2040s, or even as late as the 2070s, others suggest that such a state change might come a great deal sooner (Overland and Wang 2013). With less snow and ice cover, the darker water surface will absorb more solar radiation and contribute unwittingly to a positive feedback process of enhanced warming. As the warming continues, further dangers exist inland with regard to the thawing of the Siberian permafrost regions and the weakening of the Greenland inland ice. Both Polar Regions also contain the world's largest ice sheets, but in the circumpolar North it is the melting of the Greenland inland ice that has captured scientific and public attention in recent years as one of the starkest examples of global climate change, placing Greenland at the epicentre of processes of ecological reconfiguration. Scientific climate models suggest that average temperatures in Greenland will rise by more than 3°C this century, which would mean large-scale melting of the inland ice. Even if future conditions stabilize somewhat, a scenario where the inland ice could eventually melt completely is possible, even though it would take several centuries to do so, with sea level rise of up to seven metres (AMAP 2011).

While there is some disagreement within the scientific community as to the role of long-term natural variation in the Arctic region, most concur that the sea ice distribution is thinning due to rising temperatures. This is reaffirmed within the Arctic Climate Impact Assessment (ACIA), produced under the auspices of the inter-governmental forum the Arctic Council, and which provides an important summary of the contemporary state of affairs. As the report notes, 'Arctic average temperature has risen at almost twice the rate as the rest of the world in the past few decades. Widespread melting of glaciers and sea ice and rising permafrost temperatures present additional evidence of strong arctic warming. These changes in the Arctic provide an early indication of the environmental and societal significance of global warming' (ACIA 2004: 8).

Even if the recession of 2008–9 slowed carbon emissions, the legacies of past carbon emissions and projected future growth means that global temperatures are likely to rise regardless. The International Panel on Climate Change's (IPCC) global assessments and scenarios are important elements in determining the influence that climate change is expected to have in the Polar Regions. It is the IPCC reports, for example, that often form the basis of media reporting and political strategies pertaining to likely environmental change, especially as the ice-covered parts of the world including the Himalayan-Tibet glaciers (the so-called Third Pole) are subject to public debate. The 'safety' of the Polar Regions becomes an integral part of arguments over, for example, competing conceptions of what is the most desirable level of greenhouse gas concentrations in the atmosphere with figures ranging from 350 to 550 parts per million frequently cited. At present, it is estimated by the US government's Carbon Dioxide Information Analysis Center that the world stands at approximately 385 parts per million. When other gases such as methane are included, the overall level approaches 435 parts per million of carbon dioxide equivalent. It is highly likely that 450 parts per million will be reached by 2020.

Climate change scientists warn that we can no longer assume that our environment can be modified to suit our needs, and the prospect of global warming of about three or four degrees is not an attractive one. Without radical action to decarbonize the planet, the evolving geographies of climate change will likely as not emphasize a patchwork of places experiencing sea level rise, desertification, flooding and shifts in local and regional weather patterns. Hundreds of millions of people may well be forced to migrate as some areas simply become uninhabitable. The distinguished climate change scientist, James Hansen, has warned that the world has already passed any sensible upper limit for atmospheric concentrations of carbon dioxide, and we are heading for an overall global temperature, by the end of this century, last seen over 30 million years ago.

The Polar Regions will not only continue to foreground debates over ongoing climate change but also provide the source material needed for intelligent debates about long-term climatic change. Ice core data, for example, provide something like 800,000 years' worth of estimation of carbon dioxide concentrations and general climate patterns. What is critical is to understand better how places like the Arctic and Antarctic are bound up with calculations about probabilities, risks and uncertainties. The kinds of uncertainties that abound include future emissions, the resilience of human communities around the world, and the efficacy of decarbonization strategies. Moreover, any assessment about climate change needs to confront the fact that there remains a strand of scepticism in Arctic rim countries including Denmark and the United States. But it is just that – a strand. In the United States, the military, especially the Navy and Coast Guard, are working explicitly with the idea that the Arctic Ocean will be ice-free (at least in the summer season), and that state change is due to anthropogenic climate change. As noted above, invocations of 'dangerous

52

climate change' have been hugely significant in shaping public debates and strategies relating to what agencies and states should do in response, say, to ongoing sea ice thinning in the Arctic Ocean.

A note of caution needs to be added, however. Historically, these concerns over a changing Arctic climate are not without earlier foundation. For the last 200 years or so, scientists and political leaders have reflected on the possibility of there being an 'open polar sea', free from sea ice. What we can conclude is that this sense of the Arctic being 'disturbed' is one that is widely shared by scientists and indigenous knowledge holders who, in their different ways, record and reflect on change, whether it be the extent of sea ice, the movement of migratory species, the presence of alien or invasive species, or the ebb and flow of weather and sea conditions. This sense of disturbance is not unique to the Arctic, as scientists note how the sea ice around the polar continent is increasing rather than decreasing in the case of the northern latitudes. Stronger winds are believed to be a critical factor in ice production around Antarctica and it is believed that the 'ozone hole' might have contributed to a deepening of lows in western Antarctic and this in turn encouraged greater winds to cause ice production in areas where polynyas (open water surrounded by sea ice) are located. While acknowledging increases in sea ice cover, Antarctic scientists believe that eventually sea ice growth will reduce and that the production and retention of thinner ice will impact on sea ice coverage, in both the Antarctic and Arctic.

Perturbation for us is a major driver because it not only refers to ecological/physical disturbance but also conveys a broader sense of unsettlement; a driver that haunts the colonization of the Polar Regions as much as it informs future-based discussions about the fate of sea ice, permafrost, ice cap stability and the like. While Inuit communities have largely sought to live within the limits imposed by Arctic environments (as well as anticipating the possibilities of successful engagement with them), European and other later colonialists were eager, even determined, to colonize, to exploit, to settle and to administer lands and seas, often with disastrous results whether it be industrial pollution, species collapse due to over-exploitation or forced resettlement of northern communities in the name of sovereignty and security projects.

Amplification

Perhaps one of the most important and indeed profound changes affecting both the Polar Regions is a loss of exceptionality and a sense of both the magnification and amplification of various issues, stresses, impacts and transformations. In a physical geographical sense, amplification refers to the change in net radiation balance (e.g. warming trends) that encourages a more substantial change of temperature shift near the Polar Regions compared to a global average. According to the full scientific report of the Arctic Climate Impact Assessment (ACIA 2005), which provides a detailed account of exactly how the Arctic is warming at least twice as fast as the global average, surface and air temperatures are increasing, ice cover is melting (thus shifting the albedo balance) and oceans and seas are absorbing greater heat, especially if they not covered by an icy surface that reflects solar heat. Physical scientists have warned that the Arctic environment is vulnerable to further amplification, as the thinning of residual ice cover alongside a decline of overall sea ice extent contributes a positive feedback. Ice will continue to melt even when there are moderate melt seasons alongside evidence that snow cover in the High North is also in long-term decline (Hodgkins 2014). So marine and terrestrial environments in the Arctic are increasingly acting as energy sinks and with decreasing ice and snow cover, there is more energy available to melt permafrost and remaining sea ice.

The IPCC (2013) notes that the Arctic will continue to warm in the twenty-first century and that there is no reason to think that Arctic sea ice extent and thickness will not continue to decline. Amplification is believed to be responsible for a trend that may well result in the loss of September sea ice in the coming years. While experts disagree about timing, most accept that the Arctic Ocean will be largely ice-free in the summer season by 2050. A warming Arctic Ocean and littoral regions will have considerable implications for the future state of terrestrial and maritime ecosystems, as ocean stratification, acidification, alien species invasion, and permafrost melting make their impact felt.

What might all of this mean for the making and remaking of the Polar Regions? The persistent denudation of sea ice is leading to a world where geophysicist Henry Pollack can imagine a world without ice (Pollack 2010). Since satellite records began in 1979, this loss-making trend has been well documented in the Arctic Ocean and in July 2012 it was reported that some 97 per cent (as opposed to say 50 per cent which has been reported in the recent past) of the entire ice surface of the Greenland ice sheet was experiencing some surface melting. This was apparently quite unprecedented in more than three decades worth of scientific record keeping. It is worth recalling that the Greenland ice sheet contains some 10 per cent of the world's glacier ice mass.

While ice loss is not the same as ice-free, this reduction in ice cover has fed another kind of amplification and this is an amplification of emotional states like dread, fear and hope. We can write about these things as social scientists at times without always conveying as powerfully as we might how little things and words like 'ice free' create reactions amongst readers and listeners. How might the effects of continued physical amplification (as noted above) impact on a social-cultural amplification? On the one hand, the loss of sea ice has fed what we would describe as neo-liberal fantasies for the Polar Regions, as they are remade and reimagined as ever more accessible and exposed to exploitation. In the last decade, there have been an unprecedented number of conferences and trade summits extolling the prospects for business and corporations to seize upon such things. New geographical imaginaries such as the notion that the Northern Sea Route might be thought of as a 'Silk Road of the Arctic' rest on an appeal to excitement and hope, as diminishing physical obstacles such as ice appear to be less bothersome to those who wish to move in and out, and through the Arctic.

On the other hand, a state of fear can also be amplified. Did the planting of a Russian flag on the bottom of the central Arctic Ocean in August 2007 amplify a more fearful geopolitics? Did the image of a Russian flag provoke feelings of insecurity, unease and displeasure in a way that was different to, say, had it been a Canadian, Danish or Norwegian flag? While some political leaders were dismissive of the significance of the flag and its presence on the seabed, as we have already noted, others such as the Canadian government at the time appeared unsettled by its presence. The amplification of fear might work in a number of different ways; it might be contagious encouraging others to become more worried about other Russian activities such as military flights over the Arctic Ocean, and it might encourage a sense of 'scale-jump', whereby Russia is thought to be hell-bent on pursuing a full sovereignty agenda over a great swath of the Arctic Ocean, including the North Pole and to the south in Georgia. Thereafter, audiences outside Russia remain highly aware and attuned to the activities of the Russian government.

But the amplification of emotional states can also work in other ways. It might encourage greater scrutiny of existing governance arrangements for the Polar Regions, not only from participants who are beginning to make their influence felt such as China and South Korea but also mobilizing their knowledge to produce and promote alternative geographical understandings of the Polar Regions. This includes environmental groups such as the World Wide Fund for

Nature (WWF), indigenous peoples' organizations such as the Inuit Circumpolar Council (ICC) and energy companies such as Cairn. Finally, perhaps according a special or unique status on the Polar Regions was never very helpful in the first place, implying as it does that these areas provide some kind of comforting redemption from processes such as industrialization and resource extraction. Advocating wilderness status relies on a rather different emotional state, where the Polar Regions provoke feelings of reassuring exceptionality rather than a feeling that they might be connected to the globe in all its messiness and even nastiness.

In recent years, the way in which bodies and things are appearing to be amplified in the Polar Regions is evident in the social and environmental sciences' usage of words such as 'tipping points' and 'thresholds' in polar climates, ecosystems and societies, beyond which, it is suggested, will lie a global future shaped by dramatic, far-reaching and irreversible climatic, environmental, economic, political and social change (e.g. Wadhams 2012; Wassman and Lenton 2012; Cornell et al. 2013). In particular, the term 'tipping point' has become increasingly popular in scientific and media usage to refer to the world in which we live, to the consequences of the way we live with the world, shape and transform it, our relationships with it, and the effects of our actions. It prompts anxious discussion characterized by a nervous anticipation of a future shaped by dramatic, far-reaching, and irreversible climatic, environmental, economic and social change. It influences the development of narratives of ecological catastrophe and humanity in crisis and it is tremendously powerful in discursive, rhetorical and metaphorical senses (Nuttall 2012b).

Conclusion

We have used the title 'making and remaking' for this chapter to highlight the manner in which the Polar Regions are neither self-evident spaces nor straightforward categories. While we dealt with definitional issues, our focus here has been to think about forces and processes that help to create and recreate configurations of people, objects, practices, sites and affect. The actions of nation states, including the eight that are routinely described as 'Arctic states', act and react to a world made up of those configurations stretching across the Arctic region and well beyond. Ideas about proximity and distance, and presence and absence inform this chapter in the sense of reminding us that even those of us most distant from the Polar

Regions are connected in some manner, whether through shipping routes, environmental campaigning, travelling objects, consumption patterns, and/or solidarities with the Arctic's indigenous peoples. By focusing on making and remaking, we draw attention to the fact that there are multiple forms of the Arctic and Antarctic, and that globalization, amplification, perturbation, polarization, securitization and legalization all contribute in some form or another.

As earlier European explorers sought to accumulate knowledge about trade winds, harbours, oceanic passages and resource potential, so contemporary actors scramble to acquire a greater understanding of navigation potential, ecosystem vulnerabilities and resource availability. What we so often witness, in effect, are rival forms of geographical knowledge and the manner in which they contribute to legal, political, scientific and environmental projects. These forms of geographical knowledge interact with one another and they contradict one another. And yet they are crucial to the manner in which both the Arctic and Antarctic are represented, understood, calculated, exploited, polluted and/or governed. How we respond to places ultimately depends on how, where and why we construct places in the first instance and the kind of geographical imaginaries they help to generate and circulate.

— 3 —

UNDER ICE AND SNOW

The Arctic and Antarctic are being thought of in more explicitly three-dimensional ways, the implications of which need to be better understood. In other words, we find ourselves in a situation where enormous effort has been made over the last hundred years in striving for better understanding of what lies beneath the ice, water, permafrost, rock and snow. The onset of the Cold War in the late 1940s accelerated that curiosity and investment, as strategic planners and political leaders recognized that there was an 'ice curtain' and not just an 'iron curtain' to investigate, watch and protect. Even in the Antarctic, there was no shortage of speculation about what the Americans and Soviets might be doing under and on the ice. To know one's enemy required one to know their environments and habitats as well as you know your own.

Ice and snow fascinate and frustrate. Speculation as to what might be buried, hidden and/or lost in it remains a lucrative business. Artists and writers, scientists, explorers, bounty hunters, sailors and scientists have all found ways to 'harvest' ice and snow. As film director John Carpenter in *The Thing* (1981) recognized, you could also have fun with ice and snow. Imagine for a minute that an alien spaceship is buried beneath the Antarctic ice and then have it discovered and emancipated by curious polar scientists. Trapped in their scientific station, the alien life form mutates from husky and human until a dramatic fiery standoff with a sole human survivor. The original version of the film, *The Thing from Another World* (1951), imagined the alien object being discovered in the High Arctic by US military and scientific personnel and in this close encounter it is the alien that loses out. Carpenter's version is all together bleaker and perhaps in keeping with the worsening of global geopolitics more generally.

Fiction writers, too, have been fascinated with what lies beneath the icy worlds of the Arctic and Antarctic. Clive Cussler's *Arctic Drift* (2008) is concerned with a race to find Franklin's lost ships and the source of a mysterious silvery metal cargo they were carrying before they disappeared – as if anticipating the eventual Canadian real-life find, HMS *Erebus* is discovered with Franklin's coffin (the members of the Canadian team who located *Erebus* in September 2014 have speculated whether Franklin's casket is in the hold) and the book is replete with references to sovereignty disputes in the Northwest Passage, security and energy resources, and a rush to find mineral wealth. The final part of Peter Høeg's *Miss Smilla's Feeling for Snow*, first published in Danish in 1992 and translated into English the following year, takes place deep underground in an abandoned scientific facility constructed in a cave of ice and rock on a glaciated island off the Greenlandic coast to study and extract a meteorite. The action moves on to where the meteorite itself is located, in an immense space filled with tens of thousands of stalactites dripping down from the ceiling, some of them 'intertwined, like chains of cascading Gothic cathedrals; others are small and densely packed – pincushions of quartz' (Høeg 1993: 402). The problem is that the meteorite, 'black and motionless', is embedded in a glacial lake, the water of which is 'slightly milky with bubbles dissolving in the glacier ice', and infested with a deadly parasite. This Arctic worm, however, is also an object of desire for those who had returned to recover the meteorite (itself possibly alive), a major scientific discovery 'that represents a significant stage in the encounter between the stone, inorganic life, and higher organisms' (Høeg 1993: 404).

Cold War tensions and science in remote northern locations also provide a literary backdrop for some writers. In *Ice Hunt* (2003), James Rollins situates his narrative in an old Soviet base on an enormous ice island in the Arctic Ocean; Ice Station Grendel soon reveals its secrets of cryogenic experimentation some 50 years before, and it is not just the discovery of frozen human bodies that preoccupy the central characters – they are later soon terrorized by predatory prehistoric *ambulocetus natans* that have emerged from the deep freeze. Juris Jurjevics ventures into familiar waters with *The Trudeau Vector* (2005); something is lurking in the Arctic depths, but this time it is a virus that distinguishes this biothriller from works that deal with extraterrestrials, minerals and primordial creatures. In the Antarctic, Australian novelist Matthew Reilly's *Ice Station* (1998) imagines an icy world where long-lost spaceships and 'killer whales' lurk menacingly. The 'Cold War' may be over but in the hunt for the long-lost

UFO, old alliances are macerated as US, French and British soldiers battle it out for control of the hidden object.

The theme of icy preservation allows writers to explore anxieties over evolution and the nature and uniqueness of humanity. Ice is a substance in which microbes, parasites, human bodies, aliens, the results of Cold War experiments and unspeakable horrors can lie dormant until they are released and awakened. One of the best pieces of fiction about polar depths is H.P. Lovecraft's *At the Mountains of Madness*. First published in 1936, Lovecraft's work shows the obvious influence of Edgar Allan Poe's *The Narrative of Arthur Gordon Pym of Nantucket* (1838), but it is a superb novel about the terrors unleashed by the geological discovery of an ancient earthly or otherworldly presence. The narrator is a geologist on an expedition to Antarctica to drill and bore beneath the ice and rock. The objective of this scientific venture is to secure deep-level specimens of rock and soil mainly from the high mountains and plateau south of the Ross Sea. Drilling into a cave, a party of scientists from the expedition makes a startling discovery and from then on 'were to face a hideously amplified world of lurking horrors which nothing can erase from our emotions, and which we would refrain from sharing with mankind in general if we could' (Lovecraft 2005 [1936]: 27).

If fictional writers could imagine unspeakable horrors on a remote scientific station in the Antarctic then so could military personnel in the Arctic. US admirals could and did envisage a possible sneak attack by Soviet submarines, hiding in the Arctic pack ice, in the midst of the Cold War. During the Cold War both Soviet and American scientists were capable of imagining outrageous plans to use nuclear bombs to blow up sea ice and use nuclear engineering to clear the ice and snow from underlying rock. So while overlying ice and snow could camouflage helpfully, it could also obscure and frustrate. As the subject itself might infer, ice was a slippery proposition for Cold War military planners.

In this chapter, we look at the interest in polar interiors, sub-surface environments and ocean depths for the explicit purpose of challenging a dominant view of polar geopolitics – in other words, falling back on those horizontal imaginaries like wilderness, resource frontier and on the 'edge of the map'. We identify a series of Cold War legacies whereby the Polar Regions were subjected to ever more intense forms of earthly surveillance (Turchetti and Roberts 2014). While so much contemporary attention is placed on surface level changes such as sea ice cover and shifting wildlife distribution and habitat, there is a tendency to underplay the resilience of a vertical, even volumetric, geopolitical imagination and the way in which the ocean depths, the lands obscured by ice and snow, and even the chilly skies continue to be enrolled in geopolitical imaginaries and projects. We give credence to the importance of depth and interior geographies rather than simply a concern with surfaces, horizons, geographical areas, lines of latitude and longitude, political boundaries and borderlands, exclusive economic zones, baselines and maritime regions. We show how enduring interest in what lies under the ice and snow remains caught up in a series of scientific, resource and geopolitical scrambles for knowledge, power and access.

Going Beneath and Beyond: Thinking Volumetrically

In an era of abrupt and rapid climate change, cartographic representations of the Polar Regions showing receding and diminishing sea ice have become increasingly familiar. This visual rendering and understanding of the planet's circumpolar places remains focused on surfaces and outer layers, with maps and other images depicting the gradual loss of whiteness, the appearance of deep blue ocean space, and the northwards movement of the tree line. The Arctic Climate Impact Assessment is an example of just how powerful surface imagery and the technologies of such production can be when they are deployed to bring the Arctic into focus as a region undergoing dramatic change. Indeed, ACIA underscores Jody Berland's point that the production of satellite images 'extends a history of landscape depiction which prefers its nature at a distance' (Berland 2009: 246). In a series of high resolution colour maps, satellite images and graphics in both the comprehensive scientific assessment and the shorter distillation report of ACIA's key findings, we are invited to gaze like Apollo (Cosgrove 2001). From the top of the globe, we look (with alarm) at declining snow cover, the melt zone of the Greenland inland ice, changes in summer sea ice extent, northern places splashed with oranges and deep reds indicating rising temperatures, changes in the boundaries of discontinuous permafrost, northwards shifts in the ranges of plant animal species, alterations in atmospheric and oceanic circulation, sea level rise, and incoming, diffused and reflected solar radiation (ACIA 2004, 2005). This a view from high above of a changing Arctic represented by lines, colours and arrows. The region is visualized through the products of technologies of affect. It is intoxicating.

Despite the best attempts of social scientists involved in ACIA to draw attention to the realities and lived experience of climate change

61

for the people of the circumpolar North (e.g. Huntington and Fox 2005; Nuttall et al. 2005), the maps and graphics convey distance and detachment and give us no idea about the complexity and richness of human–environment relations, which is all the more curious given the seriousness of the melting and warming Arctic and the impacts on environment and society they aim to say something about. The projected loss of perennial sea ice in the Arctic Ocean, for example, is represented in ways that imagine northern waters in a climate-changed Arctic as increasingly blue rather than white spaces on the map, devoid of ice and transected by lines indicating transpolar shipping routes. It is a tantalizing glimpse of open water and the possibility of shorter sailing distances between Europe, North America and Asia rather than a view from the bridge of a ship navigating through northern waterways that will still be filled with ice regardless of the extent of summer melt. When ACIA's key findings were first released, they pointed to scenarios of possible environmental catastrophe; now, open polar seas hint at the Arctic's economic potential. This is reinforced through maps and graphics that attempt to make the rhetoric of anticipation legible. As Berland (2009: 247) puts it, 'maps constitute a visual index of traversable space which is pragmatically linked to the anticipated exigencies of unaccustomed, uncharted, or unfriendly movement'.

When maps of a changing Arctic do try to point to what is below the surface, they tend to mark the location, or the possibility for discovery, of oil, gas and other minerals. Ocean structures such as the Lomonosov Ridge become emboldened as and when they are connected to possible continental shelf claims. While they hint at complexity, they also reinforce a geopolitics of the surface and indicate what resources could be extracted, rather than an appreciation of the stratigraphies and structures of what is within, below and underneath. They also underplay earthly forces; the manner in which ice, snow, water pressure, atmospheric turbulence, cloud cover, seismic activity and the like have an agency that enables, slows down, frustrates and even prevents human activities and encounters. Countless explorers, scientists, pilots and miners have found to their cost how polar environments can abuse, confuse and refuse human interventions. Maps of the Arctic Ocean and the Southern Ocean rarely succeed in capturing how little we really understand of these watery environments.

Building on the work of writers such as Eyal Weizman (2003) and Stephen Graham (2004), who challenged conventional notions of geopolitics as a flat discourse and emphasized 'the politics of verticality' and 'vertical geopolitics', several geographers, including

Jeremy Crampton (2010) and Stuart Elden (2013), have pointed to understanding the complicated and multifaceted notion of territory in terms of the volumetric. Writing about securitization and aerial sovereignty against the backdrop of increasing cases in many parts of the world of violations of one state's airspace by another state's military aircraft, Alison Williams (2009) argued for the conceptualization of territory as three-dimensional volume rather than as a flat or two-dimensional area. As Elden (2013) notes, the purpose of such a starting point is to reflect further on how territory's existence is not simply taken for granted but rather seen as a mechanism for thinking about how it is filled with a volume composed variously of rock, ice, water, air, atmosphere and so on, and filled out by infrastructure including roads, bridges, tunnels, walls, mines and the like. At the same time, he argues that geopolitics 'has tended to become conflated with global politics or political geography writ large' and asks whether we could 'turn this back to thinking about land, earth, world rather than simply the global or international' (Elden 2013: 49). Territory's third dimension – the vertical with its recognition of height and depth – remains a critical accomplice for practices of geopower (Bridge 2013: 55). But, as Deleuze and Guattari remind us in their elaborations of their ideas on ecophilosophy and geophilosophy, and as Grosz prompts us in her writings on geo-power, earthly forces and shapes as well as the elemental and non-human are entangled and interfere with geopolitical projects (see Yusoff et al. 2012). Thinking volumetrically also draws our attention to anthropological studies of indigenous views of northern places as worlds of emergence, becoming and anticipation (Nuttall 2009), or as Kirsten Hastrup has it, 'how worlds emerge in a continuous process of confluence and dissociation, of movements of people and things, and of imaginaries both fixing and transcending the horizon' (Hastrup 2014: 23). This further complicates conventional – we might say excessively horizontal – geopolitical approaches to understanding the Polar Regions.

Before exploring in more depth the implication of this volumetric argument, it is worth reflecting on what is at stake. Scott (2008) calls for a reassessment of subterranean spaces as arenas of modern imperial and colonial expansion. As Elden reminds us, tunnels rarely go directly down but make use of entrance shafts and other passageways to gain access to a range of sites and resources, and he also points out that the underground 'is essentially associated with danger, risk, undermining and subterfuge' (Elden 2013: 40). Given this, thinking volumetrically about the Polar Regions, rather than simply horizontally or vertically, highlights how states and NGOs routinely use a bundle of geographical-political-legal techniques for exercising control over subterranean resources, monitoring passageways, surveying what lies within landscapes and seascapes, probing the ice-filled waters of the Arctic, carrying out remote sensing surveys of polar ice sheets and seeking to understand the bedrock, and identifying Antarctic subglacial aquatic environments using radar and subsurface measurements. The motivation for such activities was, and is, always varied and multifaceted, including paranoia, curiosity, greed, safety, hope and altruism.

The subterranean enjoys cultural and political resonance in the Polar Regions; and the Arctic and Antarctic have, and continue to attract, a coterie of imaginative and material interventions eager to probe beneath them. From Danish expeditions to Greenland in search of silver mines in the early 1600s to Henrik Rink wondering about the thickness and extent of Greenland's inland ice some 250 years later (Rink 1974 [1877]), to nineteenth-century novelists placing hollow earth theories in fictive settings and to contemporary assessments of a wealth of undiscovered hydrocarbons and scientific expeditions eager to 'open up' the subterranean worlds beneath the polar continental ice sheet, these undergrounds (and accompanying transitional spaces) are fundamental to how one might think about polar geopolitics (Griffin 2004).

Making the Ice and Snow Accessible and Legible

The onset of the Cold War is central to any analysis of the subterranean – military planners and political leaders were desperate to know more about what lay beneath the ice and snow. Military paranoia became caught up with scientific curiosity and further probing and poking of the world underneath the ice and within frigid polar seas resulted (Doel 2003; Turchetti and Roberts 2014). The Arctic and Antarctic attracted a new generation of men (there was a distinct gender regime underwriting this investigatory and explorative labour), mapping and delving into the depths of the oceans, the tundra and ice sheet to exploit, survey, secure, inhabit and exclude others from these cold environments (Bloom 1993; Collis 2009). Money came from military organizations and the Office of Naval Research was an important funder of US polar science. This naval-scientific marriage of convenience was also picked up in Cold War popular geopolitical culture as submarine movies such as *The Bedford Incident* (1965) and *Ice Station Zebra* (1968) thrilled audiences with what might lie beneath the ice and snow. The volumetric, as Bridge (2013) notes, is a primary metric of anticipation and potential, involving assessments about what a space such as the Arctic might contain and what it might become if transformed into resource and military environments. Thus, it is understandable how the figure of the 'undetected' nuclear submarine sailing under the Arctic sea ice pack became such a productive referential object in Cold War geopolitical imaginaries.

As historians and geographers of the physical and environmental sciences have reported, disciplines such as geology, geophysics, oceanography and sedimentology were vital accomplices to this volumetric strategizing (Doel 2003; Hamblin 2005). The development of Cold War era governmental power in regions such as the North American Arctic and the Russian North is a story of scientists and military planners working separately and collectively to transform the tundra and sea ice into 'vertical territory' rather than wide-open horizontal space. These interventions were also biopolitical rather than simply geopolitical (in terms of securing control). Moving people there was critical in securing the Arctic and Antarctic as volumetric territories – places to be filled. Sometimes these interventions were disastrous; the Canadian government organized relocation of Inuit to various parts of the High Arctic in the 1950s was controversial because the relocatees struggled with considerable privations and little government support in the areas in question, notwithstanding all the attempts to improve understanding of 'northern environments' (Tester and Kulchyski 1994). Sadly no amount of probing above and below the ice would have compensated for a badly thought out and funded relocation scheme.

As a consequence of what we have described, there is widespread consensus amongst historians of science that the Cold War was a major stimulus of exploratory and scientific activity in the Arctic and Antarctic, as it was in many other parts of the world in which the possibilities of confrontation between the US and the Soviet Union were reinforced by posture and rhetoric (see e.g. Dennis 2003; Farish 2010; Turchetti and Roberts 2014). To this we can add that the Polar Regions played a critical role in revolutionizing military technologies and what Paul Virilio (1989) calls the 'logistics of perception'. The Arctic in the Cold War was considered a place of potential hostility and possible invasion, a region of long-range bombers, intercontinental missiles, early warning systems, sophisticated radar and air bases. Some areas, such as northern Greenland or Russia's border with Finland, became places of dislocation for people as they were moved from their homes to make way for military infrastructure, which was

65

often accompanied by scientific facilities. Drawing on Virilio's work on battlespace, particularly as developed in *War and Cinema* (1989), we suggest that as the Arctic became strategically vital to the Soviets and to Western governments, it became a potential battlespace that not only produced a 'field of perception' but one that transformed the 'logistics of military perception'. As Virilio argued, the side that can see further and extensively, can gather more information, which can react quicker and whose reach is longer and deeper, has the potential to be the most powerful. In this way, the Arctic became vital to the sharpening of a military vision to see deeper into the territory of others, to gather information and intelligence, and to monitor and develop the capabilities to penetrate ever more intensively into that territory. Utilizing different technologies of observation and military gazing, surveillance and other forms of knowledge gathering of Arctic conditions, such as the meteorological sciences, glaciology and geology, this 'multimedia *field of vision*' (Virilio 2002) also had a volumetric dimension to it as military projects started to look within northern landscapes and beneath northern seascapes, as well as across and above them. Nick McCamley's account of the Cold War's 'secret, invisible infrastructure, the networks of underground control bunkers and radar stations stretching across continental North America and Great Britain, whose existence has never been more than a rumour' (McCamley 2002: 3) reminds us of the efforts that went into building sophisticated military bases beneath mountains and under the ice and snow.

Notable projects of such Cold War subterfuge in the Arctic include Olavsvern, a Norwegian naval base near Tromsø, designed for patrolling submarines to hide or to be resupplied in, and which extended deep into a network of mountain tunnels. Another was Project Iceworm, where the latest polar science and technology were used by the United States to excavate, mould and carve the ice sheet with the intent of concealing a system of intermediate range ballistic missiles (IRBMs) in northern Greenland (Weiss 2001; Nielson et al. 2014). In 1959–60, Project Iceworm set out to construct a nuclear-powered city – Camp Century – under the inland ice some 250 km east of Thule Air Base, as the foundation for a larger installation in which 'Iceworm' IRBMs would be housed (it was never built, although Camp Century itself was and accommodated around 225 personnel in the first few years of the 1960s). Declassified American documents have revealed the extent of the project – although Camp Century was hardly kept secret at the time, the real reason for its construction was (figure 3.1). The US claimed it was building a scientific facility and Camp Century

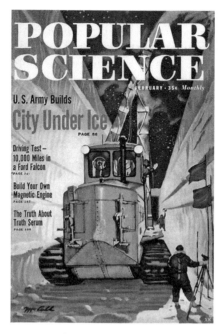

Figure 3.1 'City Under Ice' (reproduced with permission of
Popular Science)

was depicted in propaganda materials, including a film about the
mobilization of US Army Engineers to excavate and tunnel down into
the ice, as an example of the conquest of the Arctic environment and
of man's quest for knowledge (Wager 1962; Kinney 2013).

This celebratory account of the subjugation of a wild polar ice
cap and the triumph of science and technology makes no mention of
indigenous human presence in north-west Greenland and the movement and relocation of Inughuit families to make way for Thule Air
Base in the first place. Military personnel and equipment, including a
nuclear reactor to power the facility, were transported from the base
and up to and across the inland ice on trailers hauled by snow trac-
tors. Access to the inland ice was from a ramp at Camp Tuto (Thule
Take Off), a facility that had been built in 1954 to examine whether
the inland ice could be used as building material. Nielsen et al. (2014:
445) situate the construction of Camp Century within the context of
US military appropriation of northern Greenland, a process beginning
with the construction of Thule Air Base in 1951 (and possibly earlier
with military geological interest). Polar military engineering was used

67

to transform 'the snowy wasteland into a veritable city equipped with every convenience from library to warm showers' which the US Army saw as a 'stepping-stone to increased military presence on [and, quite literally, in] the icecap'. Eventually, the mobility of the inland ice crushed the tunnels, laboratories and living quarters – the achievements of American military engineering were not quite matched by knowledge of glaciological movement – but Cold War politics and difficulties in negotiating the placement of nuclear installations on Danish territory also played their part in the eventual abandonment of Camp Century (Nielsen et al. 2014). It is also salutary at this point to think that an investment of millions of dollars could also fail to deliver upon another kind of settlement project in the High Arctic.

Progress in modern upper atmosphere physics and chemistry and weather forecasting (where falling snow could obscure the skies and interfere with communication) owes much to military-supported scientific efforts during the Cold War (Harper 2008). For example, Danish research carried out in Greenland on the ionosphere became of considerable importance for US long-distance radio communications. An American military-operated network of weather stations throughout the country became critical to Denmark's meteorological ambitions to be able to imagine Greenland as part of their 'national weather map'. While Christopher Ries discusses how 'US Cold War visions of the nature of global warfare brought forth a pressing need for new kinds of knowledge about terrains and soils in northern Greenland', we can also, along with Michael Billig, note how such things feed into banal and not so banal nationalisms (Billig 1995; Ries 2012: 338).

US efforts to project and consolidate its power during and immediately following World War II were 'accompanied by an enormous expansion of military funded research, which tightened the links between science, technology and foreign policy, and affected deeply the institutions, disciplines and practices of a broad range of sciences inside – and outside – the USA' (Ries 2012: 336). The fear of a Soviet attack, even invasion, ensured that this research continued to be generously funded and supported thereafter.

But an earlier invasion arguably initiated this determination to discover more about what lay beneath the ice and snow. The Japanese invasion of Alaska in 1942 coupled with German military activity in the European Arctic brought home the importance of having accurate weather information as well as an appreciation of the distribution and thickness of Arctic sea ice. As we note further in chapter 4, as the conflict developed, US forces established a presence on Greenland

and Iceland, but Alaska also became key to the US military defence system. Air, army and navy bases were built in Alaska and the Alaska Highway through Canada was constructed in 1942. To supply the military build-up and infrastructure, the US required ready access to oil and the Canadian Oil (Canol) Project of 1943–4 involved a hastily constructed pipeline from the Norman Wells oil fields in the Northwest Territories to a newly constructed refinery in Whitehorse in the Yukon. From there, oil was transported by a network of pipelines to points along the Alaska Highway, including to a fuelling station in Skagway in south-east Alaska. Sections of pipeline were laid on the surface of the ground and crude oil frequently leaked and seeped down into the permafrost and the matter beneath. Although seen as a major event in the history of Canadian cold region engineering, it was nonetheless designed, constructed and operated with little understanding of the need for specific northern design and construction methods. It was one of the largest projects ever undertaken in northern Canada and its environmental legacy remains very much in evidence, and a testimony to what happens when you either do not understand the subterranean worlds beneath frozen soil or fail to appreciate what happens when alien substances drain and seep into it (Nuttall 2010).

In the Antarctic, German forces were active in destroying the Norwegian whaling fleet and there were fears that Japanese–German forces might invade and occupy the Falkland Islands and disrupt maritime traffic around Cape Horn. In the UK, scientists and scholars attached to the Scott Polar Research Institute at the University of Cambridge were asked to generate polar intelligence, and in the case of the Arctic required to provide information and advice on how to navigate under and above Arctic waters and negotiate the capriciousness of snow and ice. In the Antarctic, the British, fearing mounting Argentine and Chilean interest in what it considered its national polar territories, were enrolling scientists in plans to launch a secret operation designed to restore British sovereignty – to facilitate it, orders were made to collect and condense existing bodies of knowledge regarding polar environments, above and below the surface.

World War II brought to the fore how patchy knowledge was of both the terrestrial and marine environments of the Arctic and Antarctic. While systematic observations and measurements of ice, water and weather (and of indigenous peoples and the classification of material culture) were made during the first International Polar Year (IPY) in 1883–4, and although many later expeditions towards the end of the nineteenth century and into the twentieth, had increased scientific understanding of the Polar Regions – including

69

long-term programmes such as the extensive geological mapping of East Greenland begun by the Geological Survey of Greenland in 1926 – they remained frontiers of scientific investigation. For example, Fridtjof Nansen and his crew had made significant observations of the drift, motion and trajectories of sea ice in the Arctic Ocean during the *Fram* expedition in 1893–6, but knowledge of transpolar ice drift remained poor until the 1950s when Soviet and American drifting stations were deployed, accompanied later in the century by new technologies of automated weather stations and airborne and satellite observations. The first drift station on floating ice in the Arctic Ocean was set up by the Soviet Union in 1937, to be followed by extensive Soviet and US programmes from the 1950s onwards. But while little was known of the surface, and while drift stations improved understanding of sea ice, advanced meteorological knowledge and allowed for oceanographic research, even less was known about the depths of the oceans when the Soviets had set up North Pole-1, the first drift station in 1937. In the late 1930s and early 1940s, little was known, for instance, of the submerged ocean floor (Doel et al. 2006). Mapping beyond coastlines was varied and episodic with scientific-commercial investment in ocean mapping tied to sovereignty-resource agendas associated with protecting fishing and whaling interests. Terrestrial mapping was also tied to particular networks of patronage and state-sponsored interest. In the Antarctic, the UK sponsored the British Graham Land Expedition (1934–7) to enhance British mapping in the light of the growing interest of others in the Peninsula region and to improve British understanding as to whether it was connected to the polar landmass. Within two decades, however, this situation was greatly changed as ships, airplanes and snow-vehicles were put to work in ever greater numbers and intensity, probing above and beneath the ice cap and polar seas.

Given its geographical location between the two superpowers of the United States and the Soviet Union, the Arctic initially enjoyed (at least in the late 1940s) the lion's share of new investment and investigation. The US defence community was the primary driver of this activity and the Office of Naval Research (ONR) was a notable funder of data investigation, collection and analysis. Improved instrumentation, including automated depth sounders and recorders, helped oceanographers on vessels, including military submarines, to collect new information on the ocean floors, especially the Arctic and strategically significant territories such as the Bering Strait.

Military patronage was one aspect of this Cold War era enchantment of the Polar Regions. Another factor was commercial, with mining and communication corporations eager to understand the Arctic in a more volumetric manner. In the field of communications, Bell Laboratories and AT&T, when preparing to lay submarine telephone cables from Canada to Scotland alongside more covert sonic buoys, needed greater understanding of the subterranean features of the North Atlantic and North Pacific Oceans. As oceanographers increasingly appreciated, the seabed was anything but flat and rather a complex array of trenches, plateaus, seamounts, mountainous peaks and ridges. Appreciating depth and three-dimensional form was not just a commercial imperative; it was a military one. The US Navy's submarine commanders wanted to understand those topological complexities not only for their own operations but for detecting and deterring anything that their Soviet counterparts might wish to undertake. Depth soundings, as early as 1952, were caught up within an explicit national security calculation with Pentagon officials concluding that this emerging data set could not be circulated more widely (Hamblin 2005).

Seafloor sounding depended on the physical character and on seawater temperature. Rocky slopes provoked a rather different acoustic profile compared to the abyssal plain, and sea ice challenged acoustic operators in submarines. Engineers and physical scientists working with the US Navy were critical in informing advances in submarine design and technology including seabed mapping. Military planners swiftly appreciated that the Arctic Ocean was a vital space of calculation. As Soviet capabilities grew, as witnessed by their emergence as a nuclear weapon state in 1949, so the impulse to map, sound out and ultimately classify resulting data as volumetric intensified. As noted, depth data was not the only item to be affected; sea ice data, geomagnetic data, atmospheric data – almost anything that might offer further clues to those operating environments of submarines, surface vessels and airplanes.

The calculation of depth soundings proved frustrating for a new generation of oceanographers eager to investigate and map the ocean floors, and for geophysicists interested in better understanding the distribution and thickness of sea ice. Maps produced of the North Atlantic in the late 1950s, for instance, were physiographic rather than bathymetric because the US Navy did not wish this strategically important territory to be understood (at least in public) in a volumetric manner. Thinking volumetrically was something caught up, in the Cold War era, in explicit national security terms. Seafloor soundings data was more likely to be classified top secret than ever before. As Doel et al. (2006: 608) concluded, 'seafloor mapping

71

became a far more nationalistic and secretive undertaking ... the oceans [including the Arctic] became far more relevant to national security because of two technological developments: long-distance communications that utilized the acoustic properties of the sea, and antisubmarine warfare'. The emergence of new maps, however, was made possible by other circumstances with the most notable being the International Geophysical Year (IGY) 1957–8 and, in the case of seabed mapping, a commercial interest shaped by practical concerns such as minimizing costly telephone cable snagging and disintegration. From the perspective of AT&T, it made financial sense to track seafloor routes that were likely to be most conducive to long-term endurance.

Subterranean Polar Geopolitics and the International Geophysical Year (1957–8)

The 1957–8 IGY was an intense period of scientific geophysical and environmental scientific investigations of the Antarctic, the ocean floors, the atmosphere and outer space in particular (Howkins 2011). Due to geopolitical sensitivities, the Arctic was less implicated in what was dubbed a 'scientific Olympics' because both the United States and the Soviet Union wanted to do their own military-funded research on what lay beneath the ice and snow (Collis and Dodds 2008). Its timing, a worldwide affair involving 67 countries in data collection, analysis and exchange, was carefully planned. The genesis of the IGY lies with a high-level scientific conversation in April 1950 involving James Van Allen, Sydney Chapman and Lloyd Berkner, all of them distinguished geophysicists/scientists. The rationale for something that was later to be termed the IGY was military-scientific. As we have noted, there was a confluence of interest amongst the physical and environmental sciences community and military planners; both constituencies were eager to acquire a better understanding of the Earth in more volumetric terms. How did the oceans, the polar ice sheet, the tropical rainforest and the earth's weather systems interact with one another? Oceanographer Roger Revelle and geophysicist H.E. Suess were, already on the eve of the IGY, warning that the answer to such a question was likely to far exceed the national security calculations of the Cold War military planner. As they wrote, 'Human beings are now carrying out a large-scale geophysical experiment of a kind that could not have happened in the past nor be reproduced in the future' (Revelle and Suess 1957). Just as a phalanx of scientists was eager

72

to explore further the earth's geophysical properties, others were warning of the geophysical agency of the human species itself.

By 1951–2, this initial speculative conversation mutated to formal endorsement by leading scientific organizations such as the International Council of Scientific Unions. The IGY was eventually established, after much negotiation and exchange of views, for a period encompassing July 1957 to December 1958. The planning period was fraught with difficulty. As we have noted with depth soundings, collecting data that contributed to a more volumetric understanding of the earth was not straightforward. Who would benefit from a greater understanding of the ocean depths, the earth's atmosphere and the distribution and thickness of ice in the Polar Regions? How would this exploratory activity be monitored let alone circumscribed? Would interested IGY parties be given free range in spaces such as the Antarctic and outer space? Would the Soviet Union and the United States share information regarding sea ice distribution and thickness? As Peder Roberts (2014) noted, the Soviet Union was a world leader in Arctic sea ice and permafrost research, for reasons that were as much strategic as they were economic. The Soviet Union was eager to discover what lay beneath Siberian ice and snow (resources), and intent on monitoring what might reside beneath and beyond the Soviet Arctic coastline.

In Antarctica where the geopolitical stakes were different, 12 states established 48 research stations of varying sizes and locations. Many became permanent and ushered in a distinct period of human engagement with the Antarctic. The IGY precipitated an extraordinary phase of exploration and investigation but did so at a time of Cold War geopolitical tension and the onset of decolonization. This placed considerable strain on the principle that scientists should exchange freely their data with one another and that their respective governments should be committed to accepting that IGY Antarctic parties were able, on scientific grounds, to establish research stations regardless of existing sovereignty claims. The Australian government and press found this a difficult proposition to accept when it became clear that the Soviet Union was planning to establish stations in the Australian Antarctic Territory. Newspaper headlines such as 'Red flag near South Pole' (*Sydney Morning Herald* 1956) warned readers that the Soviets might establish secret underground bases and missile launch sites, which might terrorize Australian cities such as Perth and Adelaide (cited in Dodds 2002: 82). And worse still, would the Soviets leave their bases after the IGY ended?

Underscoring the IGY was a paradox. For the previous 15 years at least, the environmental and physical sciences assumed ever greater strategic salience. The military, as a funding institution, needed better understanding of the upper atmosphere (ballistic missile trajectory guidance), the ocean depths (anti-submarine warfare), underground geology (detecting covert nuclear weapon testing) and land environments such as the Arctic (infrastructural and force planning). This placed a premium on research in polar geophysics, upper atmospheric studies, oceanography, geography, geology and seismology. Earth and environmental scientists, were to varying degrees, co-opted to national security agendas, more so in the Soviet Union than in the United States and Canada. But polar scientists were in the front line of this relationship. American polar geographer Paul Siple noted in 1948 that, 'The Arctic affords a straight line of attack to the Eurasian centres of our potential enemy, and because of that, if for no other reason, we must give full consideration, to the best scientific exploitation of the Polar Regions' (Siple 1948). Within ten years, the North American Arctic and Greenland were enrolled into the IGY and used as test sites to fire and monitor rockets designed to improve understanding of the upper atmosphere. There were, in addition, two drifting ice stations locating on floating ice, maintained by the US Air Force, working on acquiring a greater understanding of natural mobilities.

The end result was to give further impetus to those who were to ascribe later the emergence of a military-industrial-academic complex in the midst of the IGY and beyond (Needell 2000). This placed considerable pressure on the ideals of scientific internationalism and perhaps explains why confidence-building measures such as 'Open Skies' policies were difficult to operationalize. Both superpowers were reluctant to share their data and knowledge of the Earth's volume with one another. There was scientific cooperation in the Antarctic but it was underscored by an anxiety about what knowledge of the polar ice sheet might facilitate. In the Arctic, Richard Powell (2008) has noted, however, that this apparent thirst for geographical and gravitational knowledge about the northern regions, in the service of ballistic missile and anti-submarine operations, galvanized the Canadian government at the time to establish and fund a Polar Continental Shelf Project (PCSP), with the expressed rationale to use scientific endeavour to bolster Canadian sovereignty. Rather than the Soviet Union looming large in this geopolitical imagination, it was a fear of the United States and what it might be doing in terms of collecting and processing ever greater quantities of scientific data about the North American Arctic. The long-range airplane and the nuclear submarine in the hands of its Cold War ally seem to afford more opportunity than before to probe above and below the icy surfaces of the Arctic. As Powell (2008) remarks, the Canadian scientists attached to the PCSP were at times ambivalent about the confluence of science, geopolitics, law and nationalism, but deeply aware that topography, cartography, geophysics and geography were integral to Cold War sovereignty and security agendas.

Sizing up the Antarctic Ice Sheet

The Antarctic was, arguably, the mainstay of the IGY's scientific priorities and yet the most geopolitically controversial. While the launch of the Soviet satellite Sputnik captured the world's imagination in October 1957 (and Soviet IGY planners divulged little to their fellow participants about their rocket and satellite plans), it was Antarctica's depths that witnessed the most transparent clash between sovereignty and security. The scientific rationale for the Antarctic dimension of the IGY was deceptively simple; scientists were aware that their understanding of the polar continent and surrounding seas was patchy. One area of speculation was the thickness of the polar ice sheet and what kind of subterranean landscapes existed underneath it. Scientists were eager to understand the area, the volume of the ice, its mass and energy, the nature of the earth beneath it, and the past, present and future history of the ice sheet and polar continent. While estimates existed for the thickness of the ice sheet, they were based on limited aerial surveying and 1930s experiments with seismic techniques (Dean et al. 2008).

Geopolitically, the polar continent remained deeply disputed. Seven countries, including the three counter-claimants Argentina, Chile and the United Kingdom, pressed particular national claims to the region. The United States and the Soviet Union rejected those claims and reserved the right to press their own on the basis of their record of exploration, discovery, mapping and, in the US case, permanent inhabitation. Following a meeting in Paris in 1955, agreement was reached that scientific research in the Antarctic could only proceed if a spirit of scientific internationalism prevailed as opposed to narrow geopolitical competition. By the time the IGY started, the 12 parties committed to Antarctic research agreed that these contending sovereignties could not be allowed to interfere with published programmes of scientific research (Dodds 2002). This meant, in effect, that claimant countries such as Argentina, Chile and the UK had to accept that other countries such as the Soviet Union and the United States were going to establish research stations in 'their' sectors. It was a tense time for all claimants, and the decision by the United States to establish a South Pole station and the Soviet Union to create a base at the Pole of Relative Inaccessibility was indicative of the intent of both superpowers to ensure that nowhere on the polar continent was 'out of bounds'.

American scientists such as Charles Bentley, who were responsible for carrying out a series of seismic traverses across the Antarctic continent, conducted one of the most notable research projects during the IGY era. From January 1957 to January 1959, a team of glaciologists and geologists dug holes into the snow and ice and detonated explosives for the purpose of 'shooting seismic' and using the resulting information to determine ice thickness and underlying subsurface geology. Building on the achievements of the earlier Norwegian-British-Swedish expedition, these seismic techniques were proving vital for polar research and were used as much in the Arctic as they were in the Antarctic by the late 1950s. The seismic traverses undertaken by Bentley and his team extended, at their longest, by some 2,600 km and by patiently braiding together the different traverses, they established for the first time a continental wide profile of the Antarctic ice sheet. It was found that earlier estimates of ice thickness (ranging from 600–1,200 m thick) were underestimates, as the US IGY team concluded that at its maximum the ice extended to around 1,200 m.

The subglacial geology was found to be even more extensive and varied than previously thought. The base surface, in the most extreme case, was found to be some 2,500 m below sea level. Later work, using airplanes and techniques such as radio echo sounding, helped to complicate further the mapping of the ice sheet and subglacial topography. The US Navy flew some 40,000 km in total and they criss-crossed Antarctica taking aerial photographs and conducting seismic surveys. The US was not alone and Soviet expeditions also contributed to an enhanced understanding of the underlying geology of the polar continent and the distribution of mountain ranges, deep valleys and subglacial lakes. This volumetric understanding of the Antarctic as a patchwork of rock, ice and snow was important not only scientifically and geopolitically (fine-tuning seismic techniques had relevance elsewhere) but also culturally in the sense of helping to redress the persistent scare stories about surviving secret German bases and UFOs emerging from Antarctic mountains and even under the ice sheet.

Some 50 years after the IGY and those early seismic traverses, the Bedmap consortium project (involving 14 countries), using satellite data, radio-echo sounding data, seismic surveys and a host of other data sources, released a map of the ice thickness and underlying geology of the polar continent (Lythe and Vaughan 2001). They estimated that the ice mass was around 26 million cubic km and that the deepest point on the continent lies some 2,800 m below sea level. Using 25 million individual survey points, it is the most detailed representation of the Antarctic ice sheet. It will contribute, so scientists believe, to enhanced understanding of the ice sheet itself, and feed into climate change modelling about the likely impact of further warming on global sea level change.

In April 2014, the Scientific Committee on Antarctic Research (SCAR) organized a four-day meeting in Queenstown, New Zealand, to develop and refine a final list of 80 priority research questions, grouped under seven themes, for future Antarctic research. One of the themes is 'Dynamic Earth – Probing Beneath Antarctic Ice', and a range of questions set out suggestions for understanding such things as how the bedrock geology under the Antarctic ice sheet informs our understanding of supercontinent assembly and break-up through earth's history, and how tectonics and dynamic topography affect the spatial patterns of sea ice change on all timescales. One of the frontiers of polar scientific exploration that has preoccupied Antarctic glaciologists for the last decade or so has been the continent's subglacial aquatic environments, the volumes of liquid water at the base of the Antarctic ice sheet several kilometres beneath the surface, which have been identified by scientists using airborne and surface radar measurements. The first observations were made more than 50 years ago while research was being carried out to determine ice surface landmarks to aid flight orientation (Robinson 1960; Siegert 2000). Almost 400 subglacial lakes have been identified since the first airborne radio-echo sounding programme in 1968–9 detected a signal attributed to water beneath the Russian Sovetskaya Station in East Antarctica, now called Vostok Station. Those signals enabled scientists to locate Lake Vostok, the largest subglacial lake yet discovered. Subsequent airborne surveys and satellite radar altimetry have revealed considerably more about the dimensions of subglacial lakes. Sealed from earth's atmosphere for millions of years, the mapping of these lakes is considered a scientific imperative because they are 'important components of the central regions of glacierised Antarctica' (Siegert 2000: 43) and are likely to contain information about microbial evolution, past Antarctic climate, the formation of ice sheets and ice sheet dynamics. As Siegert (2000: 43) puts it, 'they influence ice dynamics, act as sedimentary traps and contain a vast amount of water' and they have a major influence on the dynamics of the ice sheet above it. Significant effort has been put into drilling beneath the Antarctic ice sheet to reach some of these lakes.

While scientists can learn a considerable amount about subglacial lakes from remote sensing and ice cores, the scientific intent is to drill deep down through the ice to reach these pristine places to gather samples of water, sediments, microbial communities and the underlying rock. Three major programmes have been underway in recent years – the penetration of Lake Vostok by the Russians (who reached lake water in February 2012), drilling down to Lake Ellsworth in West Antarctica by the British and American plans to gather samples from the Whillans Ice Stream, which is also located under the ice in West Antarctica. Such exploration into the depths and possibly some of the darkest places of Antarctica has provoked a certain degree of anxiety. Despite the scientific arguments about subglacial lakes being unique laboratories found nowhere else on earth, relics from an ancient past, or ecological experiments, Russian activities raised concerns in the late 1990s and early 2000s because of the use of kerosene as a drilling fluid. Environmentalists have also expressed their worries that drilling down to subglacial lakes would contaminate pristine environments and allow for the intrusion of non-indigenous micro-organisms, and arguments have been put forward for employing strategic environmental assessments when assessing the exploration of Antarctic subglacial lakes (Lamers et al. 2014).

What is striking about all this sustained scientific interest in the Antarctic ice sheet, pre- and post-IGY, is the manner in which the Antarctic is no longer imagined to be a featureless white mass. While earlier explorers and surveyors bemoaned the disorientating landscapes of the polar plateau and administrators and governments worried about Cold War militarization descending on the Antarctic ice, we now have a far more nuanced understanding of ice thickness, ice stability (and instability) and underlying geology, including the tantalizing prospect that there might be evidence of life in the deepest and remotest parts of the planet covered by the polar ice sheet. The fate of the ice sheet is now routinely caught up in broader imaginaries and narratives about the future of the earth and scenarios based on further ice sheet instability, ice shelf collapse and sea level rise. While we continue to find evidence of novelists and film-makers speculating about the presence of secret underground worlds, lost UFOs, magical resources and alien species, the reality of our icy worlds may be more humdrum yet profound; human agency is acting with a geophysical force that early explorers would have considered unfathomable.

Measuring the Polar Seabed

About the same time the IGY was unfolding, the United Nations Convention on the Law of the Sea emerged in 1958 and led to a major transformation in the political and legal geographies of seas and oceans, including the continental shelf. While UNCLOS was revisited later in the 1970s and 1980s, the legal status of Antarctica, the high seas, the ocean floors and even outer space were all increasingly transformed by new legal agreements. A combination of continued exploitation, exploration and geopolitical tension radically altered human awareness and understanding of subterranean geology and bathymetry. As with other areas of the earth, the role of the Cold War was a major driver in stimulating funding and activity around the world's seas and oceans (Collis and Dodds 2008), and as historian of science Helen Rozwadowski notes, in the US context, the ocean was being actively imagined as a frontier environment. As she concluded, 'Before World War II, most Americans thought about the ocean, if they did at all, as the source for seafood and a surface for steamship travel, shipping, or warfare. Submarine warfare particularly attracted attention seaward and ensured massive federal investment in marine science and technology to improve understanding of the ocean, especially the depths. After the war, the metaphor of the frontier became attached to the ocean' (Rozwadowski 2012: 579).

The ocean and the seabed were also legal 'frontiers' of sorts. Changes in international maritime law were incentivizing new investment in marine research as coastal states were given 'sovereign rights' for the purpose of exploring, exploiting, conserving and managing the resources of the seabed and subsoil. Initially, 'sovereign rights' applied only to the natural prolongation under the sea – in other words the geological continental margin. If the coastal state in question did not possess a natural prolongation, then no continental shelf rights would prevail. This changed with the negotiations encompassing UNCLOS III (1973–82) because all coastal states were deemed to enjoy 'sovereign rights' over the seabed, regardless of whether there was evidence of natural prolongation. A legally constructed continental shelf replaced a geophysical one. Irrespective of geological connections, it was proposed that the coastal state enjoys up to 200 nautical miles of continental shelf rights and this consolidated further the concept of the exclusive economic zone.

Under Articles 76 and 77, coastal states enjoyed the possibility of asserting their sovereign rights over wider areas of the seabed and subsoil if they could demonstrate to a technical body, the United

Nations Commission on the Limits of the Continental Shelf (CLCS) that their continental shelves extended further. Criteria were established so that it was possible to argue that continental shelf rights might apply some 350 nautical miles and even beyond from the coastal baseline depending on seabed thickness and relationship to ocean depth. While the scientific data in question can be expensive and cumbersome to collect, it provided a mechanism for coastal states to colonize new spaces. We might even term it as a sort of volumetric expansionism.

While this extended continental shelf provision did not attract immediate attention from coastal states, the Russian Federation was the first to submit its relevant materials to the CLCS in 2001. They hoped the CLCS would issue a favourable 'recommendation' on its assertion that it enjoyed enhanced sovereign rights in the outer continental shelf regions off Russia, especially in the Arctic. At the time, this submission drew little public attention but this changed radically in 2007 when media reports were filled with stories that the Russians were attempting to claim the North Pole and large swathes of the Arctic Ocean itself. The actual process of data collection of the geology of the outer continental shelf undertaken by Russian scientists was one that followed the procedures set out in UNCLOS and followed on from an initial ruling from the CLCS that asked the Russian Federation to provide new scientific materials for the central Arctic Ocean.

The reality of the situation was that the flag-planting incident itself, while evocative of previous exploratory feats involving flags such as the conquest of Everest and the Moon, was not legally significant. What was far more important, if less visually appealing, was the scientific work. The rules governing outer continental shelves and the establishment of sovereign rights are technical and largely understandable to only a select group of international lawyers and geophysical scientists. In essence, any coastal state wishing to extend their sovereign rights has to demonstrate, geologically and oceanographically, that it is connected to that continental shelf – there are formulae to be used and it involves detailed surveying of the seabed itself, seabed thickness and the relationship between the continental shelf and other features such as submarine ridges.

It is an expensive and time-consuming business, and the CLCS requested that the Russian authorities resubmit their materials for the Arctic portion of their continental shelf even if they were content to issue recommendations for the Barents Sea region. Other Arctic states, namely Canada and Denmark, are accumulating their own scientific data for the purpose of establishing their own outer continental shelf submissions. Under the terms of UNCLOS, coastal states muster their information about the depth and extent of extended continental shelves and submit it to the CLCS. All three of the above Arctic states want to map and to extend the borders of their national territories and all three have aspirations regarding the North Pole and establishing the volumetric limits of their northern territories.

These calculations designed to make the seabed legible both to coastal governments and the CLCS will take years to resolve. In 2014, the CLCS had received 75 submissions from coastal states, two revised submissions and made about 20 recommendations, including ones for Norway and Denmark with respect to the continental shelves around Svalbard and the Faroe Islands, respectively. Legal and technical arguments will continue as coastal state submissions are slowly evaluated by the CLCS and any eventual resolution of sovereign rights over the Arctic and Antarctic seabed will depend upon bilateral and multilateral negotiation depending on where the outer limits of the Canadian, Greenlandic and Russian continental shelves are established in the central Arctic Ocean.

The situation of the Antarctic continental shelf is even more complicated than the Arctic. Seven claimant states believe that they are in effect coastal states but the international community at large contests those sovereignty rights and the Antarctic Treaty places a premium on questions of territorial sovereignty being suspended for its duration. This has not stopped some claimant states such as Argentina from making a direct submission in 2009 to the CLCS and asking it to make a recommendation on its sovereign rights to the continental shelf areas contained within its sovereign claim – the Argentine Antarctic Territory. Others have protested and asked the CLCS not to make such a recommendation. The CLCS is unlikely to adjudicate on that particular matter, mindful of the deeply contested nature of the Antarctic Peninsula and the provisions of the 1959 Antarctic Treaty.

In other parts of the Antarctic continental shelf, the Australian submission of 2004 raised the intriguing prospect of sovereign rights over the continental shelves off Heard and Macquarie Islands in the Southern Ocean extending into the area of application of the Antarctic Treaty. What would be the implications for the Antarctic Treaty and other international legal instruments such as the Madrid Protocol if Australia approved mining and mineral exploitation on the extended continental shelf off those subantarctic islands? In 2008, the CLCS approved the vast bulk of the Australian submission and the extended continental shelves of Australia's subantarctic islands such as Heard were duly defined. As with other areas of the Southern Ocean, the presence of islands to the north of the Antarctic Treaty's area of application (i.e. south of 60°S) provides further opportunities for the sovereign rights of coastal states such as Australia, South Africa, UK, France and New Zealand to extend closer to or indeed within the area of application.

The measuring, mapping and evaluating of the polar seabed have been hugely productive of geopolitics. Lines have been redrawn and they go across as well as downwards. While it is not clear at all whether states will ever exploit the resources contained within and on their extended continental shelves, the geometrics involved with continental shelf submissions tell us something about the kind of calculations coastal states are prepared to make when it comes to mapping and evaluating the remotest of all their territories, including the surface and subsoil of the polar seabed. Other claimants such as New Zealand and the UK reserve the right to make a future outer continental shelf submission regarding their Antarctic territories, confirming what many assume to be a 'holding position' in case the Antarctic Treaty System were to collapse in the future.

Toxic Legacies and Polar Futures

Attention to the volumetric is also important – more immediately in the North, but also for future discussions of Antarctic resources – because of the way the Arctic is being reimagined and represented as a new frontier for oil, gas and mineral extraction, a theme we consider in depth in chapter 5. With the region increasingly described as a place of possibility, a resource space in which there are supposed to be enormous opportunities and prospects for developing extractive industries to supply global energy needs and meet increasing global consumption demands, questions of subsurface governance, and of subsurface interests and technologies of discovery, research, extraction, production and consumption, come to the fore. How various interests in and activities affecting the subsurface should be governed is increasingly complicated by the emergence of what are seen as non-conventional techniques such as hydraulic fracturing, which re-engineer the ecosystems of below in order to bring trapped gas to the surface. And it is crucial to bear in mind that the histories of relations between states and indigenous peoples and other northern residents have often been shaped by a concern with exploration and ownership of the subsurface. The subsurface has also been at the heart of protracted discussions relating to treaty making, land claims and self-government in many parts of the North.

It is difficult to find places in the North where indigenous people and local communities do not have a history of experiencing the economic, environmental and social impacts of extractive industries, whether directly because of proximity to sites of exploration and industrial activity, as cumulative effects, or in the form of trans-boundary or long-range pollutants, which leave their traces beneath the surface as well as above in the bodies of residents of indigenous communities. If, from the perspective of industry, the circumpolar North as a resource frontier is one of its most distinctive attributes, it is also the case that many northern regions can be more accurately described as extractive peripheries, in which resource dependency and a flow of benefits to the core have persisted (Nuttall 2010). The materialities of many historical mining operations, for example, live on in community memories or in the depths of toxic material in con-taminated ground and polluted lakes and rivers, lurking within and leaking slowly to the surface and entering human bodies, fish and animals. Water, fish, animals, as well as people themselves, also give toxic material an agency as substances that are mobile and passed on and inherited. As they do so, they compound the dreams and plans for 'clean' extractions from the polar volume and remind us that there are always multiple agencies at play here and there.

In the Canadian North, Arn Keeling and John Sandlos describe Sahtu Dene concerns with the environmental and health effects of radium and uranium mining between the 1930s and early 1960s around Great Bear Lake in the Northwest Territories in terms of envi-ronmental justice, and point to a number of government–community studies that 'highlighted the deep psychological scars remaining for many of the generation who lived through this period and who associate the mine with the painful social and economic changes associated with industrial development. For people who benefited little from the resource wealth [and who were] removed from their territory and who [continue to] live with the environmental legacies of mining, a sense of bitterness and injustice remains' (Keeling and Sandlos 2009: 118). In their discussion of the legacy of the North Rankin nickel mine at Rankin Inlet in Nunavut, Cater and Keeling (2013) discuss how tailings and other toxic elements are felt by residents to have a persistent and malign presence in the landscape. At the same time, while they are concerned about the environmen-tal legacies of the mine, people engage with the mining past which

UNDER ICE AND SNOW

stimulates forms of remembering of the positive aspects of the days of mining activity. Equipment, machinery, old buildings and mine shafts are repositories of community memories, significant for identity and for thinking about a future that is also likely to be characterized by resource extraction. As they put it, 'the material remnants of mineral extraction link past events and present encounters through ongoing and everyday encounters, as people come into contact with material objects' (Cater and Keeling 2013: 64).

In his discussion of the forgotten uranium mine of Paukkajanvaara in Finnish Karelia, Colpært (2006) illustrates how the rehabilitation and reclamation of old mine sites 'raise questions of dimensionality alongside political economic concerns' (Elden 2013: 47). With attention focused on the Arctic as a new frontier for oil, gas and minerals, it is often forgotten that circumpolar lands bear witness to a history of leaks and seepages from the extractive industries that are residual at the surface and deep underground in volumetric environments bifurcated by shafts, mines and tunnels. While we can see the abandoned harbours, roads, airstrips, sheds, machinery and tailings, we can trace if we wish those objects and substances that lurk within the rock and ice. Abandoned mines such as the North Rankin nickel mine and many more, such as the lead zinc mine at Maarmorilik in northern Greenland or the Innatsiaq (Josva) copper mine and Amitsoq graphite mine in south Greenland, are examples of numerous sites around the circumpolar North in which people have probed and dug deep beneath the surface to extract materials for transformation into global commodities.

In recent years, companies have begun to explore the prospects for uranium mining in Greenland, either directly or as a by-product of rare earth minerals extraction, with the Australian company Greenland Minerals and Energy focusing on potential development of both uranium and rare earths at Kuannersuit (known as Kvanefjeld in Danish), a plateau near the south coast town of Narsaq (which has a population of around 1,200). Yet while Greenland Minerals and Energy has been active there since 2007, the prospects of uranium mining at Kuannersuit have been subject to speculation for 60 years. The area lies within what geologists call the Ilimaussaq intrusion. Mesoproterozoic in age and significant for the presence of a variety of rare minerals, the uranium mineralization at Kuannersuit has been investigated since 1955, when Niels Bohr, then chair of the Danish Atomic Energy Commission, initiated radiometric exploration of the area. Drilling and mapping programmes followed and 180 tons of ore were mined in 1962. The dream was that Greenlandic uranium would

84

be extracted and used in Danish power stations to make Denmark independent of oil.[1] The Kvanefjeld Uranium Project was launched in 1978, financed by the Danish Ministry of Energy. Feasibility and environmental impact studies were carried out in the early 1980s, a 1,000 metre-long tunnel was dug into Kuannersuit, and some 10,000 tonnes of ore were extracted until the project ended in 1982. Although the environmental impact assessment concluded that there was a considerable lack of knowledge of both the technical aspects of the exploitation of uranium and of the environmental character-istics of the region, the assessments and feasibility studies overall were positive for future work to be developed, but the adoption of a zero-tolerance policy meant Denmark would no longer be a potential buyer of the uranium.[2] In October 2013, Greenland's parliament voted by a narrow margin of 15–14 to lift the zero-tolerance policy on mining uranium and other radioactive materials, which had been in place since the Danish and Greenlandic parliaments had agreed on it in the late 1980s. Since then it has been thought of as an impedi-ment to a potentially lucrative development in south Greenland that some political and business elites have long wished for. The uranium issue in Greenland is controversial, provoking spirited public debate and protest and raising questions over Danish–Greenlandic relations and decision making over its extraction and export. At the heart of concerns are the lack of public participation in discussions about extractive industries and the future of Greenland (Nuttall 2013, 2015), and it has also become an international issue, as environmen-tal organizations expressed their concern about Greenland becom-ing a producer and exporter of uranium. The story of uranium in Greenland, however, is also one about how people perceive and relate to subsurface resources and how the histories of probing and explo-ration have been a longer part of Greenland's political history than recent controversies over extractive industries would have us think.

Conclusion

The Cold War made clear that both the US and the Soviet Union desired to put the Polar Regions to work – militaries and politicians wanted to get under the ice and snow. Both superpowers wanted to

---

[1] Nielsen, "Portræt af et fjeld" ("Portrait of a mountain").
[2] Pilegaard, *Preliminary environmental impact statement for the Kvanefjeld uranium mine*; Kalvig, *Preliminary mining assessment of the uranium resource at Kvanefjeld*; Nielsen, "Portræt af et fjeld".

know what lay beneath in order to further their geopolitical objectives. Resource evaluation, maritime awareness, scientific curiosity and technological experimentation were emblematic of why the volume and volumetric mattered. In the Arctic, various sites were also used to test nuclear weapons (and dump nuclear materials), to test personnel and equipment, to see whether it was possible to live under the ice let alone establish long-term secret bases in the ice, amongst the immense boreal forestry and hidden within mountain ranges. For those things to be possible, the Polar Regions were subject to environmental and physical scientific scrutiny, aided and abetted by military funding and support and new bodies of international maritime law, which incentivized coastal states to map and evaluate their subterranean territories.

Drawing attention to the volumetric also requires us to ponder the influence that humans have as agents of geological and geophysical change in the Anthropocene, evident in the modification of the environment through the excavation of rock, soil, sediment transfer, the movement of earth and the creation of artificial ground (Price et al. 2011). It should also provoke us to think further about the role of the non-human. Snow, ice and rock encourages, facilitates, prevents and frustrates human projects, including those eager to colonize, exploit and nationalize the Polar Regions. Those projects eager to discover the thickness of the Antarctic ice sheet and evaluate the shape and thickness of the polar seabed were ones that depended on the interaction between human and non-human agents. And as countless scientists, pilots and miners have discovered, polar environments can and do offer formidable challenges, even if the presence of snow and ice can also offer opportunities, as indigenous hunters, submarine captains and ice road truckers might testify, albeit with different experiences.

We might also ponder how polar volumes can both conceal and reveal human and non-human impacts on the Polar Regions. Just take ice as an example. Things can be buried in ice. Things can seep into ice. Things can be drilled into ice. And things can be revealed as ice melts. Polar warming has revealed old artefacts and remains, highlighting how climates and environments have changed over millennia. The extraction of ancient ice volumes reminds us about how human activities in the past are revealed in the present. Roman-era metal smelting in southern Europe 2,000 years ago left a trace in Arctic ice packs. And it is not just ice that carries those traces. The bodies of indigenous peoples continue to act as living volumes, inhabited by those leaks and seepages of Cold War industrialization and military scheming.

86

——— 4 ———

GOVERNING THE ARCTIC
AND ANTARCTIC

From the Antarctic Treaty to indigenous land claims and self-government settlements in the Canadian North, governance brings into sharp relief spatial understandings of the Polar Regions. These create and, over time, sustain power-geometries – themselves indicative of the multi-scalar connections that constitute the Arctic and Antarctic. It demands attentiveness to what we term 'topographic' (i.e. fixed locations and spaces and geographical distinctions such as land/sea) and 'topological' (i.e. relationships and networks across space) arrangements. The governance of the Arctic and Antarctic has been shaped by actors and interests *rooted* in place as well by other actors, objects, practices and circuits of knowledge that have made their impact felt even if their origins lie far beyond the Arctic and Antarctic.

This complements our discussion about thinking volumetrically. For all the desire to understand the geophysical qualities of the Polar Regions in the midst of the Cold War, there came with it a growing appreciation of both the resource potential of and legal connection to other apparently remote spaces such as the deep seabed and the high seas. International legal developments such as UNCLOS, from the 1950s onwards, transformed public and political understanding and appreciation of governance challenges. They had the effect, however, of elevating and at times isolating. Laws were made, scientific knowledge was gathered, political arguments mustered and post-colonial transformation brought into play new actors and new attitudes towards areas of the planet that were considered by others to be 'global spaces' rather than nationalized or internationalized by a self-selecting few.

The governance of the Antarctic and Arctic is a more contested affair than say 50 years ago. The privileged status enjoyed by science and scientists to imagine and influence management of these parts of the world is rather less assured in the case of the polar South. Antarctic Treaty Consultative Parties argue, more explicitly than ever before, about fisheries science, conservation policies, environmental restrictions, base location, biological prospecting and resource-led futures. The social and institutional power of consensus and secrecy is also less assured than it once was when a select group of states and their statesmen (sic) discussed the merits or otherwise of conventions and protocols. As a considerable body of work emerging from the Arctic also shows, awareness of the limits of scientific resource management has, in recent decades in particular, focused interest in the environmental knowledge of the Arctic's indigenous peoples as well as other northerners, such as Norwegian whalers, Icelandic fishers and Finnish reindeer herders, whose livelihoods are based on the utilization of wildlife and domesticated animals (e.g. Huntington 1992; Anderson and Nuttall 2004; Kalland and Sejersen 2005; Heikkinen et al. 2010). Evidence from practice suggests that the application of indigenous and local environmental knowledge in development projects, environmental management and environmental impact assessment both enhances the likelihood of success of those initiatives and processes and acknowledges that northern residents are environmental experts who possess and have access to information and knowledge that is unavailable or denied to scientists, with implications for the governance of northern homelands.

This chapter explores the geopolitical and environmental complexities of the governance of the polar oceans and seas, as with other parts of the Arctic and Antarctic environments. These spaces are not simply abstract ones to be encountered around the conference table and in the corridors of power. The Arctic and Antarctic/Southern Ocean are experiential, material and representational spaces. The stories that we tell and listen to about these spaces and the way in which we engage with the Polar Regions have implications for their governance and for the geographical imaginaries that inform such governance. The physical encounters with sea ice and sea currents (e.g. polar front zone), the circulating presence of flora and fauna including charismatic animals (e.g. whales and polar bears), the deployment of metaphors and analogies (e.g. the Arctic Ocean as a polar Mediterranean, the Southern Ocean as a last frontier, the Antarctic as common heritage), the historical legacies of exploration and exploitation (e.g. the search for oceanic passages and the hunt for marine mammals) and the affective allure of vast spaces (e.g. the high Southern Ocean and central Arctic

Ocean as inviting fishing grounds and mineral hotspots) all play their part in shaping governance regimes.

Making 'Global Antarctica'

Until the middle decades of the twentieth century, an increasing number of nations became interested in Antarctica and they undertook expeditions of geographical and scientific exploration. Governments began to lay claim to parts of the continent alongside plans to expand and deepen resource exploitation and as such created new relationships and networks. Britain was the first to do so in 1908, followed later by France, New Zealand, Australia, Norway, Argentina and Chile (Britain, Argentina and Chile each lay claim to sovereignty over the Antarctic Peninsula). Scientific activity also intensified – notably following the end of World War II and especially after the IGY of 1957–8 – with the establishment and maintenance of permanent national Antarctic science programmes by countries such as the UK, the US, Russia, Norway, France, Australia and New Zealand, all with the aim of giving some legitimation to territorial claims and/or, in the case of the US and the Soviet Union, preserving the possibility of making such claims in the future (Dodds 2012).

In retrospect, the IGY was the culmination of a decade-long period in which a number of proposals were put forward by interested parties such as Chile, New Zealand and the US addressing the future governance of Antarctica. Various suggestions were posited; a condominium of the seven claimant states and the United States, a 'sovereignty freeze' for the purpose of pursing international scientific cooperation, and even some discussion in New Zealand of potentially relinquishing sovereignty claims if a new international body to manage the Far South might materialize. Overlapping sovereignty claims between Britain, Argentina and Chile provided the urgency of such proposals (figure 4.1).

The Antarctic as space was increasingly being understood topologically; as something that was connected in multiple ways to those interested parties. No longer a remote and barely understood landmass somewhere over the horizon, it was being incorporated into the public cultures of claimant states such as Argentina and Chile, conceptualized as a site of Cold War intrigue and possibility by the United States, and imagined by the Soviet Union as a place in which it had to remind the US and its allies that it was not going to be 'excluded' from any governance discussions. As we discussed in chapter 3,

Figure 4.1 The British Polar Vessels, HMS *Endurance* and RRS *James Clark Ross*, Leith Harbour (reproduced with permission of the British Antarctic Survey)

scientists were also playing their part in reimagining the polar continent as intimately connected to planet earth and the atmosphere and beyond. Moreover, if imagined as a mineral resource frontier, this added further piquancy to the contention that the Antarctic mattered, albeit differently, to all the IGY polar parties. Relative proximity and absolute distance was not going to prevent the United States and the Soviet Union from expressing, and indeed demanding, that their interests be accounted for.

First signed in 1959 by those 12 nations active in the Antarctic portion of the IGY (Argentina, Australia, Belgium, Chile, France, Japan, New Zealand, Norway, South Africa, the Soviet Union, the United Kingdom and the United States), the Antarctic Treaty entered into force in 1961. It set aside Antarctica for peaceful purposes, suspended territorial claims and extolled international collaboration in science. Since then, 38 other nations have joined what later became the Antarctic Treaty System (ATS), full membership of which is limited to those 29 states that 'demonstrate substantial scientific interest' in the continent. This caveat was important at the time and thereafter. Treaty signatories, attentive to Cold War schisms and uncertainties, were eager to ensure that they were able to act as scientific 'gate-keepers' and thus regulate its membership. While it was never articulated explicitly, there was an underlying concern that some African states and Asian countries such as India, who had twice raised the 'Question of Antarctica' in the United Nations General Assembly in 1956 and 1958, were eager to become more involved in the management of the polar continent. In reality, the scientific criteria were highly politicized, as was apparent in the 1980s, when India and China were rapidly accepted as Antarctic Treaty Consultative Parties, whereas West Germany was made to wait far longer for accreditation.

India's intervention in the 1950s, however, raised a different kind of possibility for Antarctica (Howkins 2008). Reimagined as a global object of interest, the polar continent and its future was tied to broader debates about North–South relations and a Cold War geopolitical order, which presumed so-called Third World states should choose one side or another. The Antarctic, and later spaces such as the deep seabed and moon, became places from which to initiate debates about common heritage, rights and access and the very nature of global governance itself. Just as scientists tied Antarctica into global debates about the atmosphere, the ocean and the planet in its entirety, so lawyers and political leaders introduced another kind of braiding; a legal and geopolitical assemblage which tied the fate of Antarctica and the oceans into a wider conversation about the rights of near and distant others, regardless of their political, historical and ecological connections.

Revisiting the admission policies of the ATS, however, was essential to its long-term survival. By the 1980s, it was considered vital for it to be more inclusive, and increasingly more open as to how it conducted its political and scientific business. Debates about common heritage combined with the signing of UNCLOS in 1982 brought the future of Antarctica to the fore. Rising resource interest, moreover, in terms of Southern Ocean fishing and future prospects of mineral resource exploitation (however remote), placed further pressure to be seen to be more inclusive and more accountable. But it also ensured that Antarctica was becoming ever more 'global' in the sense of being connected, embedded and networked with a greater range of interests and actors, including distant water fishing fleets and Third World countries with little apparent Antarctic provenance.

A diverse range of countries representing developed and developing nations, claimant and non-claimant states, and nations that were ideologically opposed to one another were increasingly coalescing under the banner of the Antarctic Treaty and associated legal instruments with an apparent common goal: to use Antarctica to make 'substantial contributions to scientific knowledge resulting from international cooperation in scientific investigation' (e.g. Elliott 1994). But that common goal was never straightforward and never free from controversy precisely because those topological relationships with Antarctica varied in scope and intensity.

The rising number of countries involved in Antarctic politics and the corresponding increase in the scientific activities conducted there has led to concern over the management of human activities, and how best to avoid impacts on its environment, while at the same time allowing for its use for a 'global good'. Extractive industries in the Antarctic are prohibited by the Protocol on Environmental Protection and its Article 7, which was signed in 1991 and entered into force in 1998. Despite designating Antarctica as a 'natural reserve, devoted to peace and science', it could be argued that the protocol is an instrument that has had the effect – whether intentional or not – of not only protecting the Antarctic and enhancing the Antarctic Treaty regime, but protecting the territorial claims of the seven states namely Argentina, Australia, Chile, France, New Zealand, Norway and the United Kingdom. Political factors continue to be highly significant in the development of national Antarctic science programmes, including research station location. King George Island in the Antarctic Peninsula region has, for example, Argentine, Brazilian, Chilean,

Chinese, Ecuadorian, Peruvian, Polish, Russian, South Korean, Uruguayan and US scientists stationed there. It is a moot point as to whether such a concentration of scientists and scientific programmes on one particular island is sensible. In one sense, making Antarctica more 'global' has led to research station over-concentration and apparent duplication of research effort.

'Newer countries' to the Antarctic Treaty System such as China, South Korea, Pakistan and India have found their plans to establish bases scrutinized in a way that, as we explain in chapter 6, might reveal a form of 'polar Orientalism', whereby others such as the claimant states like Australia remain suspicious of their motives and interests in Antarctica. Despite this, however, Australia and China have a history of collaboration in logistics and Antarctic diplomacy, reaffirmed by the signing of a memorandum of understanding in November 2014. This also affirmed a commitment to the ATS, to Antarctic science, and to non-militarization and environmental protection. Some scholars have repeatedly noted, however, that there is plenty of discussion within China of Antarctica's resource potential, including its mineral resources, and this raises the possibility that senior Chinese figures imagine the Antarctic as just another resource base within a global matrix (e.g. Brady 2012). While China respects the current prohibition on mining, as noted in the 1991 Protocol on Environmental Protection, there is interest in the possibility that such a prohibition might be rescinded. This is not judged by any observers to be likely for some decades, but increased funding for scientific programmes and research stations could be seen as indicators of China's long-term investment in Antarctica and some may speculate that China might respect a particular vision of the Antarctic as a space for scientific research and environmental protection while making plans to reimagine the Antarctic as a resource frontier, albeit one already in play through fishing.

Twenty years ago, Lorraine Elliott (1994: 2) argued that environmental-scientific regimes such as the ATS are 'biased in that the hierarchy of values that underpins them privileges political and economic interests over environmental ones'. In Antarctica and elsewhere, she argued, rules for environmental protection 'are often inadequate to the task identified for them because they are compromised by other interests'. Compliance is also minimal (or perhaps not scrutinized as it could be) because there are no strong or enforceable sanctions. This is not to say that legal instruments such as the Protocol on Environmental Protection are not filled with 'rules' and procedures; rather it is perhaps to focus our attention on how things get implemented, evaluated and monitored. Whatever the good intention of scientists and administrators, the contested status of Antarctica means that the announcement, let alone the environmental impact assessment, of say a proposed research station or a marine protected area (MPA) is always caught up in a matrix of geopolitical-scientific-legal-environmental factors. For those wishing to expand their presence on the Antarctic continent such as India, China and South Korea, there remains a residual suspicion that others use environmental concerns to cast doubt on the wisdom of that expansion. For those eager to promote new MPAs, there are others who think that environmental stewardship is being used opportunistically to promote sovereignty agendas or even restrict the activities of others such as those with fishing interests such as China, South Korea, Russia and Spain.

While increasing scientific knowledge has provided an argument for continuous activity in Antarctica, along with the presence of permanent bases and scientific stations, growing awareness of the impacts of climate change and suspicions that many states are interested in the resource potential of the continent, mean that scientific activities and political motives for being involved in Antarctic research are increasingly coming under a more global scrutiny. Environmental groups such as Greenpeace and the Antarctic and Southern Ocean Coalition (ASOC) remain at the forefront of that scrutinizing role, as demonstrated by their ongoing campaigning relating initially to mineral resource negotiations in the 1980s and more latterly in areas like the establishment of marine protected areas in the Ross Sea and other areas of the Southern Ocean. ASOC has drawn public attention to the perceived intransigence of states such as Russia and Ukraine regarding living resource conservation and their interests in the fishing industry. However, as before, picking on China or Russia might feed into a 'polar Orientalism' that underplays the fact that all the states party to the ATS – especially claimant states – have their own political, economic, legal, stewardship and scientific agendas. One example is the role that Australia, in a convenient partnership with the Sea Shepherd Society, has played in challenging Japanese whaling. While Sea Shepherd is committed to stopping 'scientific whaling' in the Southern Ocean, Australian governments have been eager to use environmental stewardship to promote their own interests, and to ensure that the fate of the whale becomes an Australian concern given the proximity of whaling to 'its' waters. Whales have been drawn closer to the everyday lives of Australian citizens by the legal challenges brought by campaigners in the Australian courts and later the International Court of Justice in The Hague (Rothwell 2013).

As the recent controversies regarding whales and whaling demonstrates, Antarctic governance is complicit with competing global imaginaries. Positioned by some as a continent with unique treaty-based arrangements, where science and international cooperation can prevail, it has nonetheless been seen in radically different ways. Accused of being comprised of elitist, exclusive and anti-Third World countries, the Antarctic Treaty System has also been enrolled in different projects, some of which witnessed the involvement of the United Nations General Assembly in the 1980s. As resource interest grew alongside international legal developments such as UNCLOS, so Antarctica became globalized in a different manner and one rather more attentive to access, control and management as opposed to scientific exchange and collaboration on 'the ice'. Pulled, stretched and squeezed, Antarctica was reassembled and reimagined not as an exclusive 'continent for science' or 'frozen laboratory' but as a resource space and a site for competing ideas about the 'rational use' of resources and environmental stewardship. Australia and Japan fundamentally disagree, for example, not only about whaling as an industry but also about the whale itself and what it represents in its own right and its relationship to humankind.

Indigenous Homelands, Land Claims and Self-Government

Some governance systems in the Polar Regions refer primarily to wildlife and resource use (e.g. polar bears and caribou, whaling and fisheries), or more specifically to aspects of the environment. In the Arctic, contemporary controversies about the predominance of science in environmental policy and environmental management contexts and decision-making processes have centred around arguments for a shift from the types of top-down management of activities, technologies and resources by government or regional institutions and organizations that prevail across the region to bottom-up approaches that are inclusive of stakeholder/community-based management and governance, where local people are recognized as legitimate experts with rights to participate in decision making and in the implementation and administration of lands, waters, wildlife and non-renewable resources. Diverse groups of stakeholders and rights holders (local communities, indigenous peoples' organizations, NGOs, etc.) have become involved in dialogues with institutional/official experts (scientists, policymakers) about the assumptions and practices of science, and the validity of local knowledge (Passelac-Ross and Potes 2007; Wright 2014). This is different from the Antarctic, where the absence of indigenous human populations emboldens the words and work of scientists and environmental campaigners as they seek to 'root' their views about Antarctic futures in assessments about non-human population numbers and distributions.

Given this, it is important to make a distinction between government and governance. We could say, simply, that government refers to specific kinds of public and state institutions vested with authority by the state to make decisions on behalf of an entire community, country or nation, whereas governance, while including the institutions of government, also encompasses other social forms, practices, institutions and non-governmental organizations that play a role in decision making and in implementing and overseeing the arrangements that emerge from such dialogue and processes of inclusion. In a sense, governance encompasses both formal and informal dimensions of decision making, and involves civil society. Writing about institutional arrangements in Arctic indigenous societies, Fondahl and Irlbacher-Fox describe governance as

> the exercise of legitimate authority within a group to make decisions regarding the allocation of resources and the coordination and management of communal and, to some extent, individual activities. The term refers to the principles, institutions and practices that a collective employs to regulate relations among its members, and between its members and the external world. Governance stipulates how resources are shared and managed; it guides social relations. It is informed by endogenous norms and practices mediating such relations. (Fondahl and Irlbacher-Fox 2009: 2)

The ability to govern and make authoritative decisions about the use of the environment and its resources, and of social practices, 'depends on the knowledge of one's environment, and the demonstration of this knowledge through skilled practices' (Fondahl and Irlbacher-Fox 2009: 6). Leaders in small Arctic communities, by way of example, particularly in hunting and fishing societies, often emerge because of their considerable knowledge of animals and the environment, and their notable success as hunters and providers – such as in Greenland, where the *piniartorssuaq* (great hunter) is someone acknowledged for their skill, knowledge, prowess and experience, and often consulted by others about things that affect the community (Nuttall 1992).

Across the Arctic, lands, waters and resources are often at the heart of indigenous claims for self-government. Land claims submissions to governments articulate how customary resource use practices and the relationships people have with the environment and animals remain important for maintaining social relationships and cultural identity. A large body of anthropological work also chronicles in ethnographic detail how such practices and relationships nurture and define individual, family and community identities and reinforce and celebrate the relationships between indigenous peoples, animals and the environment (e.g. Nuttall 1992; Anderson 2000; Dahl 2000; Caulfield 2007). Hunting, herding, fishing and gathering activities, for example, are based on such relationships. They link people inextricably to their histories and ancestral environments, their present cultural and ecological settings, and provide a way forward for thinking about sustainable livelihoods (Nuttall et al. 2005). Relations between indigenous peoples and the nation states of the Arctic are complex, entangled and often fraught. In Alaska, Canada, Greenland, northern Fennoscandia and northern Russia, there are compelling similarities in the long-lasting effects of colonialism and in the ways that the social and cultural transformations resulting from state intervention in the lives of indigenous peoples have had significant implications for relations between people, animals and the environment (e.g. Nadasdy 2003; Anderson and Nuttall 2004; Kulchyski and Tester 2007). John Rennie Short has argued that relations between indigenous peoples and colonists/settlers in North America need to be understood as cartographic encounters in that, rather than Europeans discovering and exploring a new world, successful European mapping and settlement was the result of an exchange of information between newcomers and indigenous peoples, but 'over the long term these cartographic encounters gave the newcomers enough knowledge and power to render superfluous the indigenous people' (Short 2009: 13). To this we can add that colonial encounters also involved practices of erasure which not only ignored indigenous conceptualizations of place as settlers and developers imagined and appropriated northern lands as spaces of unbounded wilderness or as frontiers, but which removed and resettled indigenous people. In western and north-west Canada, for instance, new forms of property and land use were introduced as a result of the signing of treaties that not only re-conceptualized ideas of space and place, but made way for the expansion of agriculture, the movement of settler societies, and the marginalization of indigenous people. Indigenous activism and political movements began in several parts of the Arctic in the 1960s and 1970s to address these disruptions and shape arguments for land claims settlements and forms of self-government. These arguments rested in part on

97

the articulation and expression of indigenous human–environment relations, the importance of animals, and ideas of indigenous identities and homelands that distinguished indigenous peoples from more recent settlers and sojourners in the Arctic. Essential to this, at least in Alaska, Canada and northern Fennoscandia, has been a mapping of indigenous use of lands, waters, rivers, lakes, ice and place names, a process that challenges the dominant geographies and cartographies of Arctic states.

In the 1970s the landmark Inuit Land Use and Occupancy Project was an extensive mapping of historic and contemporary Inuit movement, dwelling and perceptions of the environment in northern Canada and provided a baseline for the negotiations of the Nunavut Land Claims Agreement. It was a process of counter-mapping, indigenous cartography and gathering indigenous knowledge of Arctic environments (Milton Freeman Research Ltd 1976). Remapping the Arctic often expresses how, from an indigenous perspective, the environment is understood to be dynamic and complex – nature is not separate or independent from society. As is often expressed by indigenous peoples in the Arctic and beyond, human beings, animals, and all aspects of the world emanate from the same creative source. Such expression is found from the Arctic to the tropics – on the tundra, in forests, deserts, and in steppe environments (e.g. Nuttall 1992; Descola 1994). A growing body of anthropological work over the last few decades, as well as by indigenous activists and scholars, shows that indigenous perceptions of the environment and people's relationships with the environment must be understood 'in terms of engagement, practical experience and perceptual knowledge' (Rival 2005). Such engagement, experience and knowledge also go some way to form the basis of people's social relations and interaction. Often, though, this is not understood by the state, industry and others who seek to use and appropriate indigenous lands.

Situations where indigenous communities come into conflict with industry and development projects, such as oil, gas and mining, or with logging and hydropower development, are far too numerous. In 2001 the Inter-American Court of Human Rights ruled that Nicaragua had violated the property rights of the Mayagna Awas Tingni by granting permission to a foreign company to carry out logging operations in forests on the community's traditional lands (Inter-American Court of Human Rights 2001). Community leaders argued that the Nicaraguan state had failed to demarcate communal land and to take the necessary measures to protect indigenous property rights over ancestral lands and natural resources. The court noted that Nicaragua's actions violated the American Convention on Human Rights, which recognizes and protects the property rights of indigenous peoples. Jentoft (2003) argues that this case has similarities with a situation in northern Norway. There, the Norwegian government argued that a valley important for Saami reindeer grazing and for fishing and hunting was *terra nullius* and therefore effectively subject to the control and management of the state. However, the local Saami community counter-argued that, because they had always used the land, they had long-established occupancy and user rights – in effect, the land was theirs. The case went to the Norwegian Supreme Court, which supported the Saami position.

Local communities are thus often forced to articulate the importance of the land in both traditional and contemporary ways to ensure the protection of important cultural, economic and natural areas and species. In some cases, as in the examples above, indigenous peoples fight legal battles in the highest courts or in international tribunals. In others, communities aim to deal directly with industry by engaging companies in dialogue. By way of illustration, in 2005 the Gwich'in of Canada's Northwest Territories worked to gather local indigenous knowledge about land near the route of a proposed natural gas pipeline, which was a large part of the major Mackenzie Gas Project in the Mackenzie Valley region. While not overwhelmingly opposed to the pipeline – indeed, many community members expressed hopes that economic benefits would result from its development – their aim was for the project proponents to recognize that protection of certain lands from industrial development is necessary for the continuation of Gwich'in cultural practices and traditional livelihoods. During community workshops held prior to public hearings on the pipeline project, for example, elders talked about a mossy area directly on the proposed pipeline route that is known for cloudberries, an important cultural and nutritional resource. Within this area, portions that are swampy are also good for muskeg tea and blueberries. For the energy companies there was nothing immediately significant about this marshy area, but it would be crossed directly by the pipeline and was within a proposed access route for gravel development, which would be located on a high hill. This hill is known as Oo'in, meaning 'look-out site' in Gwich'in. During the workshops, elders pointed out and emphasized the importance of the site as a place to scan the low-lying land and keep a watch for migrating caribou, as well as its probable importance as an archaeological and ancestral site. The pipeline project proponents listened to community concerns, heard about local knowledge of the land, the enduring and irreplaceable nature of animals and plants for people's food security, and, where possible, took all this into account during the planning of the pipeline route (Nuttall 2010).

Such respectful dialogue appears to be a rare occurrence despite the proclamations of resource companies to be committed to corporate social responsibility and nurturing strong and respectful industry–community relations, as McCreary and Milligan (2014) discuss with reference to the ontological politics of an encounter between a pipeline company and indigenous peoples in northern British Columbia. The Northern Gateway Project will take crude oil produced in the oil sands region of northern Alberta and move it through a pipeline from just north of Edmonton to the British Columbia coast. McCreary and Milligan examine the struggles of the Carrier Sekani people in the permitting process and the emerging politics of indigenous resistance – they show with considerable insight how the legal obligations to include traditional knowledge actually work to prevent processes of indigenous becoming from being considered in public hearings and the regulatory framework. Similarly, in Greenland local perceptions of and relations with land, waters and animals, and of local understandings of the environment as being in a perpetual state of becoming are ignored in the social and environmental impact assessments that are carried out for mining projects and for the seismic surveys conducted for hydrocarbon exploration – local experience and knowledge are erased by the production of technical knowledge and in political and industry discourses about Greenlandic frontier environments and subsurface resources (Nuttall 2013).

One of the most strident assertions of indigenous rights movements is that indigenous peoples are distinct political and cultural communities, and that movements promote ways for indigenous peoples to regain and exert control over their lives and lands, illustrated by land claims and campaigns and struggles for self-government. For indigenous peoples, the relationships between people and land are indissoluble and so it follows that land rights are inextricably linked to land claims. Arctic peoples share with other indigenous peoples around the globe the historical experience of having lost, or faced the threat of losing, their prior rights to their lands, territories and resources as a result of colonization. Land claims movements – and hence the assertion and reclaiming of land rights – are struggles for self-determination, self-government, autonomy and justice. Moreover, the protection of indigenous lands, the recognition of rights to lands and resources, and the settlement of land claims are often seen as essential for the very survival of indigenous peoples and cultures. The right to land is thus seen as inseparable from the right to life, and separation and expulsion from the land is often seen as 'ethnocide' and 'cultural genocide' (Barume 2010). Colonization processes have seldom recognized the rights of indigenous peoples to land – indeed, colonizing powers fail to recognize that traditional lands are homelands for indigenous communities and Cold War geopolitical imaginaries were more likely to imagine the Arctic as either a testing ground for military technologies or an area that needed to be incorporated into a world geodetic system, designed to improve the accuracy of intercontinental ballistic missiles (Cloud 2002; Pike 2010). Land was, and often still is, seen as empty (or worse a kind of 'waste space') by colonizers and settlers including Cold War military planners. This classification of indigenous environments as sacrifice zones (Kuletz 1998) 'violates fundamental territorial and cultural rights and aspirations of indigenous peoples' most notably with initiatives such as Project Chariot, which proposed to re-engineer Arctic landscapes through the use of nuclear explosions in late 1950s Alaska (Dahl 1993: 103).

Self-determination does not in itself mean that indigenous peoples have effective control over resources, but self-government is about being granted sufficient rights to practise autonomy effectively. Some Arctic states have recognized the claims of indigenous peoples for land rights and self-government and a number of significant and transformative settlements have been negotiated and reached over the past three to four decades. Notable among these settlements are the Alaska Native Claims Settlement Act (1971), Greenland Home Rule (1979) and Greenland Self-Rule (2009) and, in Canada, the James Bay and Northern Quebec Agreement (1975–7), the Inuvialuit Final Agreement (1984) and the Nunavut Land Claims Agreement (1992). Dahl (1993: 108) has pointed out the importance of recognizing that, in making these claims, the Arctic's indigenous peoples such as the Inuit, Dene and Athapaskans, have not demanded the creation of autonomous ethnic states – they have attempted to assert rights as distinctive peoples within nation states, and to be recognized as having such rights, even if political independence has been a goal of some political parties in Greenland, for example.

Recognizing Land Rights and Rootedness: Examples from Canada

Over the last 30 years or so, land claims and self-government negotiations between the Government of Canada and Inuit and First Nations have resulted in the formal and legal recognition of Aboriginal rights, specifically rights to land, including rights to subsurface resources in some cases. The map of parts of Canada has been radically altered by the settlement and implementation of what are known as comprehensive land claims, particularly in the Far North where there were outstanding claims to Aboriginal title over traditional lands. In most cases, these settlements and new arrangements have transformed the political, social and economic lives of indigenous peoples because, as Bone (2003: 192) has observed, 'They provide the means and structure to participate in Canada's economy and society and yet retain a presence in their traditional economy and society.'

Land claims processes – in effect, modern treaties between indigenous people and the state – are unique to North America in the northern circumpolar regions and, in Canada at least, are a first step towards self-government. As Irlbacher-Fox (2005: 1152) states, the negotiating processes are 'compelled by legally recognized aboriginal rights to lands and resources'. The Inuvialuit of the western Arctic reached a land claim in 1984, the Gwich'in did so in 1992, the Sahtu Dene in 1994, and the self-governing territory of Nunavut emerged in 1999, following the 1992 Inuit land claims agreement. In the Northwest Territories, the Tlicho (Dogrib) First Nation signed a land claims agreement with the Canadian government in 2003; negotiations for land, resource and self-government rights continue with the Dehcho First Nations and the Akaitcho Dene; while negotiations for self-government are in progress with the Inuvialuit, Gwich'in, and the Sahtu Dene community of Deline. In Yukon Territory, the Umbrella Final Agreement – reached in 1988 and finalized in 1990 – is the framework or template for individual land claims agreements with each of the 14 Yukon First Nations; ten First Nations have signed and ratified an agreement; another two have signed agreements which were not ratified after being defeated in referendums; and two are still being negotiated. In many parts of what in Canada is called the provincial North, i.e. the northern areas of the provinces of Alberta, British Columbia, Saskatchewan, Manitoba, Ontario, and Newfoundland and Labrador, treaties signed between indigenous groups and the Crown in the late nineteenth and early twentieth centuries remain in effect. The Inuit and Cree of northern Quebec signed the first modern treaty, the James Bay and Northern Quebec Agreement, with the Canadian government in 1975, while the Inuit of Labrador signed the Labrador Inuit Land Claims Agreement in 2005.

The 1984 Inuvialuit Final Agreement extinguished Inuvialuit rights and interests in land in exchange for ownership of 91,000 km^2 of land, cash compensation of C$170 million, preferential hunting rights, participation in resource management, subsurface mineral rights to a small area of land, and a provision for future self-government. In 1992 the Gwich'in Comprehensive Claim Agreement gave the Gwich'in ownership of 22,331 km² of traditional lands with subsurface mineral rights to one-third of that area. Other rights and benefits include C$75 million, a share of Mackenzie Valley resource royalties, participation in the planning and management of land, water and resource use, and a federal commitment to negotiate self-government. The Sahtu Dene and Métis Agreement (1994) gave beneficiaries 41,437 km² of land, with subsurface rights over 1,813 km², and it contains similar provisions to the Gwich'in agreement. The Nunavut Final Agreement of 1993 (which preceded the Nunavut Act establishing the government of the new Nunavut Territory) established fee simple title for approximately 21,000 Inuit beneficiaries to just over 18 per cent of the total territory of Nunavut, which includes mineral rights to over 36,000 km².

By providing cash compensation and setting resource royalty levels, these agreements have established administrative structures and provided financial resources to make it possible for Aboriginal communities to participate effectively in the mainstream Canadian economy, and in so doing mark a way forward for the possibility of reducing inequalities and eliminating social exclusion. They have defined and recognized entitlements to lands and resources (and their use), and have included guarantees to Aboriginal peoples for specific access to natural resources, including subsurface minerals. They have also initiated co-management regimes, which are forms of shared governance, for decision making over natural resources, land-use planning, wildlife management and environmental issues (Bone 2009), although some of these regimes often still have their shortcomings (Nadasdy 2003; Anderson and Nuttall 2004). Unlike earlier agreements and treaties, which emphasized the exchange of lands for various forms of compensation, and which redefined land use and property ownership, comprehensive land claims agreements in Canada's North have emphasized instead the importance of land and resource governance over land sales (Irlbacher-Fox 2005: 1152).

With comprehensive land claims, Aboriginal rights in Canada have by and large become more clearly defined. Significantly, as was the case in Alaska with the Alaska Native Claims Settlement Act, the negotiation over land claims and rights over use and access to resources occurred in the face of plans for megaproject development, particularly oil and gas projects. Irlbacher-Fox (2005: 1152) observes correctly that, while they have been misunderstood as taking an anti-development stance, 'One of the basic goals of indigenous peoples has been to participate in and control aspects of development – to engage in it rather than be excluded from it.' Land claims are symbolic for indigenous peoples in the ways they acknowledge recognition of their rights as well as provide them with a means to ensure economic, social and cultural survival. Importantly, land claims, as well as other legislation, have also meant that those wishing to develop lands and resources in Canada's North must recognize Aboriginal title to lands and resources and the duty to consult.

Over the last few decades, as the essence of this book explores, the Arctic has become a focus of global attention concerning the future of its peoples, wildlife and environments. Climate change is only one of the many challenges faced by the Arctic's peoples and ecosystems. Lichen – the main winter staple of reindeer – and other tundra plants have high levels of contaminants from pollutants emitted from factories far to the south. The Chernobyl nuclear reactor explosion in 1986 had a severe impact on Saami livelihoods in north-west Russia and northern Fennoscandia. Fallout from the explosion included radioactive caesium, plutonium and strontium which contaminated water, fish, animals and tundra soils. Polychlorinated biphenyls (oily, human-made substances known more popularly as PCBs, which, for example, evaporate from rubbish dumps and burning oil) have been found in the breast milk of Canadian Inuit women at rates four times higher than in women in southern Canada. Ultraviolet-B (UVB) radiation penetrating the thinning ozone layer has had damaging effects on human skin, eyes and the immune system.

Inuit worry about the toxic nature of seal blubber because high levels of atmospheric and marine pollution have entered the Arctic food chain. These persistent organic pollutants (POPs) break down and decompose far more slowly in the Arctic than in warmer regions of the world, and so pose greater dangers to humans and animals alike. PCBs, for example, have been proven to cause cancer and damage the hormonal and neurological development of children. There are also alarmingly high concentrations of PCBs in some seal, walrus and polar bear populations which threaten the animals' reproduction. High levels of mercury have been found in the liver tissue of ringed and bearded seals, both of which are eaten in quantity by Inuit and constitute the primary source of food for polar bears. In Greenland, one in six persons has dangerously high levels of mercury in their blood, while other toxic chemicals found in Inuit include toxaphene and chlordane. Tundra and marine ecosystems in the

Arctic are also at risk from the dumping of nuclear waste and heavy metal contamination. Nuclear test explosions have been carried out near Novaya Zemlya in the Russian Arctic, with the resulting radio-activity affecting the northern part of the North Atlantic and the Barents Sea. When tourists set off to travel in high latitudes, their journeys, far from being adventures in a remote, timeless, pristine wilderness, criss-cross contaminated ground and navigate through polluted waters.

So, while there has been much progress in recognizing and indeed cementing the rights of indigenous peoples in the North American Arctic, topographical and topological relations with other parts of national territories and the wider world have made their impact on the bodies and environments of the region. Southern constituen-cies, especially during the Cold War, treated Arctic spaces as testing grounds and waste spaces, and often trampled over the interests and wishes of communities, especially in the Soviet Union. From farther afield, Arctic peoples were drawn ever closer to the efflu-ent of others, the resource ambitions of extraterritorial parties and the accidents, indifference and mishaps of European and North American societies most vividly represented by nuclear accidents and dumping.

Making the 'Global Arctic'

Work by geographer Carina Keskitalo (2004) has resulted in fine-grained analyses of how Arctic discourses have contributed to the construction of the Arctic as a region that required governance. The conceptualization of the Arctic as a region required discursive and material work, especially in the aftermath of the Cold War (see below). But more generally, she reminds us that the Polar Regions and their governance required initiative, attentiveness and collaboration in order for international bodies such as the Antarctic Treaty System and the Arctic Council to endure. This was particularly so in the case of the Arctic that had borne the industrial and strategic brunt of the Cold War in comparison to the Antarctic. In the process, these bodies and various stakeholders, both state and non-state, helped to create distinctly 'Arctic' issues or, as Franklyn Griffiths noted, the so-called 'ice states' possessed 'special responsibilities' for the Arctic region (Griffiths 1999: 11).

Over the last 25 years or so, increased concern over Arctic environ-mental change, the status of its wildlife, and the health of its peoples, has led to significant international cooperation on environmental protection and sustainable development and significant discursive investment in the Arctic as a distinct region of the international political system. Before the late 1980s, geopolitical tension between the superpowers in the latter half of the twentieth century hindered discussion on cross-border conservation and socio-economic concerns. The Arctic has seen few wars or battles that are part of wider territorial or global conflict comparative to other regions of the planet – before the twentieth century most conflicts witnessed in the far northern parts of the planet would have been local skirmishes between neighbouring communities or different indigenous and local groups, or between indigenous people and explorers, whalers and fishers. In 1606, for example, the crew of the *Hopewell*, captained by John Knight on a voyage along the coast of Labrador in search of the Northwest Passage, came under attack from a group of Inuit. Twenty years earlier, John Davis had lost four men along the same coast from arrows fired from Inuit bows. Inuit oral history is replete with accounts of warfare within and between different groups that predate the arrival of Europeans. Some of this may have been driven by retribution, but protection of hunting grounds and local territory as well as food scarcity and long-standing animosity between groups may also have been reasons for conflict.

Yet some northern lands and seas have witnessed bloodshed on a large scale or have been on the brink of major conflict as a result of state-endorsed military campaigns. Battles have been fought between countries over northern lands and waters of strategic importance, and it is this aspect of the geopolitical value of the North which has gone some considerable way in defining and influencing relations between regions and states. The Karelian isthmus at the eastern end of the Baltic Sea, including the area around Lake Lagoda, has long been important for trade between Europe and Asia and tensions simmered between Finland and Russia over this area following Finland's declaration of independence in 1917. In the late 1930s, the eastern Baltic was in focus because the Soviets, concerned with Nazi Germany's expansionist ambitions, wanted a naval base on the northern shore of the Gulf of Finland. On 30 November 1939, the Soviet Union invaded Finland and fighting occurred in Finnish–Russian border regions, spreading north of the Arctic Circle. The Winter War, as it is known in Finland, became an epic struggle for Karelia, with the Soviets aiming to conquer Finland overall. The Finns held out longer than the Soviet Union had expected, until they signed the Moscow Peace Treaty in March 1940, ceding 9 per cent of Finland's territory, including much of Karelia, to the Soviets (Edwards 2006).

Later on, during World War II, Norway experienced the pressures of German attempts to advance north, while the northern convoy route from the United Kingdom to Murmansk on the Kola Peninsula became one of the most important means of resupplying the Soviets. The North Atlantic air route between the US and Europe also meant that Greenland and Iceland assumed new strategic significance as way points for bomber squadrons and other air reinforcements. In summer 1941, following the signing of a defence agreement between Denmark and the US, the Americans began construction of the Bluie West One airfield on a glacial moraine in the area known as Narsarsuaq in south Greenland. Other American airbases were built in Greenland – the US in fact maintained 17 military installations around the island in total – but Bluie West One saw much of the major activity. During the war, around 10,000 aircraft landed there en route to Europe and North Africa and some 4,000 US service-men were based in Narsarsuaq at the peak of the base's operations. It was often said that whoever controlled the North Atlantic would determine the outcome of the war, and this was also dependent to a considerable extent on meteorological knowledge. Accurate weather forecasts were vital for merchant shipping, the movement of naval fleets and for aircraft, and so land-based weather stations in Greenland and Iceland were important to the Allies for the data they provided. This gave the Allies an advantage over the Germans who relied on weather ships and U-boats operating in the North Atlantic as well as aircraft fitted with special weather-recording equipment. The Germans attempted to establish weather stations on Greenland's east coast to gather crucial meteorological data to help with U-boat movement and they were successful in erecting an automatic weather station in northern Labrador in 1943. On the other side of the Arctic, Japan seized and occupied the Aleutian islands of Kiska and Attu in June 1942 and it was a year later before US troops managed to evict them. Fear of further Japanese incursions into these North Pacific islands led the US and Canada to develop and construct extensive military networks and infrastructure in Alaska and Yukon, including the Alaska Highway.

For almost half a century, from the end of World War II, much of the Arctic had been a militarized zone. The Cold War effectively divided the region into two sectors – the Soviet Arctic and the Western Arctic. The speed of technological development meant that geogra-phy no longer mattered. The expanding range of intercontinental missiles and reports of submarines prowling under the pack ice made Arctic nations aware that the shortest distance between them lay over the Arctic Ocean and national perspectives on the Arctic were dominated by questions of security. Conflict – or the fear of it – influenced decision-making processes as the region emerged as a zone of potential hostile confrontation. Advances in military technology and capability allowed Arctic states to focus on nuclear deterrence. A series of radar stations were installed across the North American Arctic and the Distant Early Warning (DEW) Line was constructed across the northern parts of Alaska, Canada and Greenland in the late 1950s. It was a complex system put in place to warn of possible attack by the Soviet Union and it gradually evolved into the North Warning System in the 1980s. Substantial military build-up also happened in the northern parts of the Soviet Union, especially on the Kola Peninsula. The strategic roles played by Iceland and Greenland were redefined in the late 1940s and early 1950s as important links between the United States and Europe, but they were also considered prime strategic locations for aerial bases. Iceland became a member of NATO in 1949 and US forces, having left at the end of World War II, returned to the base at Keflavík. Iceland's geostrategic positioning in the late 1940s was explored in *The Atom Station*, the 1948 novel by Halldór Laxness, in which local associations and citizen action groups petitioned the Icelandic government 'not to sell the country; not to hand over their sovereignty; not to let foreigners build themselves an atom station here for use in an atomic war' (Laxness 2004 [1948]: 59). Laxness touched the raw nerve of Icelandic sensitivity to foreign influence at a time of newly-acquired autonomy (Iceland became independent from Denmark in 1944) – '"Why do I want to sell the country?" said the Prime Minister. "Because my conscience tells me to," he said, and here he lifted three fingers of his left hand. "What is Iceland for the Icelanders? Nothing. Only the West matters for the North. We live for the West; we die for the West; one West. Small nation? – dirt. The East shall be wiped out. The dollar shall stand"' (Laxness 2004 [1948]: 61).

The 1941 defence agreement between the United States and Denmark was renewed in 1951. The US had come to view Greenland as an important base for its bombers to launch attacks on Russia's large cities, military complexes and industrial centres and so Thule Air Base was constructed in 1951 and expanded in 1953. The base, however, was built in the middle of the traditional and culturally important hunting territory of the Inughuit who lived in and around the settlement of Uummannaq. Construction and expansion of the base disrupted the local ecosystem, affected hunting and fishing activities and necessitated the forced relocation of 27 families 140 km north to the newly-created town of Qaanaaq in 1953 – in fact the 116 people who were moved amounted to one-third of the total Inughuit population. Indigenous rights activists have argued that the resettlement infringed the defence agreement, broke international law and resulted in the unlawful confiscation of property (Brøsted and Fægteborg 1995). While disagreements persist in Greenland today over the presence of Thule Air Base, it remains an important intercontinental missile early warning site.

We can probably identify the turning point in contemporary Arctic international relations as happening in 1987, when President Mikhail Gorbachev of the Soviet Union gave a speech in Murmansk in which he called for the Arctic to be recognized as a zone for peace, and proposed greater international cooperation among Arctic countries. Significantly, Gorbachev reimagined the Arctic as a place where the northern land territories of Europe, Asia and North America converged around the Arctic Ocean. His conception of the Arctic was both terrestrial and maritime, and one that understood the region as a meeting place and not just something to be passed through, above and below the water surface. As he noted, 'The Arctic is not just the Arctic Ocean but also the northern tips of three continents – Europe, Asia and America. It is the place where the Euro-Asian, North American and Asia-Pacific regions meet, where the frontiers come close to one another and the interests of states belong to mutually opposed military blocs and non-aligned ones cross' (Gorbachev 1987).

Finland took the lead in pursuing such cooperation on a formal level as a means of addressing, along with other pressing issues, environmental problems caused by Soviet mining and other industrial operations close to northern Finland. The outcome of this was the Arctic Environmental Protection Strategy (AEPS), adopted in 1991 at a ministerial conference of the eight Arctic countries, held in Rovaniemi. The US, Canada, Denmark/Greenland, Iceland, Norway, Sweden, Finland and Russia signed the Rovaniemi Declaration and established five objectives for the AEPS: (i) to protect the Arctic ecosystem including humans; (ii) to provide for the protection, enhancement and restoration of environmental quality and the sustainable utilization of natural resources, including their use by local populations and indigenous peoples in the Arctic; (iii) to recognize and, to the extent possible, seek to accommodate the traditional and cultural needs, values and practices of the indigenous peoples as determined by themselves, related to the protection of the Arctic environment; (iv) to review regularly the state of the Arctic environment; (v) to identify, reduce and, as a final goal, eliminate pollution.

Ministerial Declarations issued at conferences in 1993, 1996 and 1997 governed activities under the AEPS. The AEPS was an unprecedented international initiative focusing on issues of environmental protection, most notably on the conservation of flora and fauna and the assessment of pollution and contaminants in the Arctic. Some Arctic countries argued that more formal, state-driven arrangements were needed to facilitate and strengthen Arctic international cooperation and to promote sustainable development across the region, mindful of growing attention from those living and working outside the Arctic. At a Ministerial Conference in Ottawa, the eight Arctic countries signed the Declaration on the Establishment of the Arctic Council on 19 September 1996 following two years of negotiations. The declaration established the Arctic Council as a high-level forum to promote cooperation, coordination and interaction among the eight Arctic states, with the involvement of the Arctic indigenous communities and other Arctic inhabitants on common Arctic issues, in particular issues of sustainable development and environmental protection in the Arctic. The chairmanship of the Arctic Council rotates for terms of two years.

The Arctic Council subsumed the working groups established under the AEPS, has developed its own working groups and launched its own initiatives, most notably the Arctic Climate Impact Assessment, the Arctic Human Development Report, the Arctic Resilience Report and the Arctic Marine Shipping Assessment, major assessments of oil and gas activities and surveys of living conditions in Arctic regions. The reports resulting from these and other Arctic Council activities provide a broad picture of the kinds of environmental, societal and health situations, challenges and needs in all eight Arctic countries, and helps guide the work of governments in assessing options for environmental protection, conservation and sustainable development. More recently, the Arctic states signed up to legally binding agreements on search and rescue and oil spill management, which also further consolidate and contribute to Arctic regional discourses and practices, including the positioning of the Arctic 8 as guardians and stewards of the circumpolar North while at the same time recognizing that global attention on the Arctic was mounting in the form of newer actors such as China, Singapore and South Korea (who became observers of the Arctic Council in May 2013).

Indigenous Peoples and Arctic Environmental Cooperation

Significantly, and crucially, the Arctic Council has recognized the importance of indigenous issues in the Arctic, and six indigenous peoples' organizations (Inuit Circumpolar Council, Saami Council, the Russian Association of Indigenous Peoples of the North [RAIPON], Arctic Athabaskan Council, Gwich'in Council International and the Aleut International Association) have status in the council as Permanent Participants. Since the 1970s, indigenous peoples' organizations have become increasingly important actors in Arctic environmental politics, giving a greater voice to indigenous peoples throughout the circumpolar North and arguing the case for the inclusion of their knowledge, experience and perspectives in strategies for environmental governance, wildlife management and sustainable development (Nuttall 1998; Shadian 2014). In the Arctic Council, indigenous peoples' organizations have made a place for themselves at the vanguard of Arctic environmental protection and sustainable development for indigenous communities and have assumed roles as major players on the stage of international circumpolar diplomacy and policymaking concerning Arctic futures.

In Greenland, Canada and Alaska, regional Inuit organizations and governance institutions have been putting into practice their own environmental strategies and policies related to resource use, and to ensure participatory approaches between indigenous peoples, scientists and policymakers to sustainable resource management and development. From an Inuit perspective, threats to wildlife and the environment do not come from hunting, as animal-rights groups and some environmental organizations would argue, but from airborne and seaborne pollutants entering the Arctic from industrial areas far to the south, the impacts of global climate change, and the extraction of non-renewable resources, such as oil, gas and minerals, and other global processes that challenge human–environment relations. In recent years Inuit have sought ways to identify, highlight and counteract such threats and devise strategies for environmental protection and sustainable development. By so doing, they claim their right to be recognized internationally as resource conservationists. This approach has been made more effective through the activities of the Inuit Circumpolar Council (ICC).

As a pan-Arctic indigenous peoples' organization representing the rights of Inuit in Greenland, Canada, Alaska and Siberia, the origins of the ICC lie in a conference of Inuit leaders held in Barrow, Alaska in 1977. Its foundation was partly in response to increased oil and gas exploration and development in the Arctic, especially the kind of development occurring on the North Slope of Alaska, planned for northern Canada and anticipated for Greenlandic waters. The ICC has had non-governmental status at the United Nations since 1983. It has challenged the policies of governments, multinational corporations and environmentalists, arguing that protection of the Arctic environment and its resources must recognize indigenous rights, indigenous environmental knowledge, and accord with Inuit tradition and culture. Since its formation, the ICC has sought to establish its own Arctic policies, combining indigenous environmental knowledge and Inuit concerns about future development with ethical and practical guidelines for human activity in the Arctic. In September 2008 in Kuujjuaq, Nunavik, the ICC organized an Inuit Leaders' Summit which resulted in a 'Circumpolar Inuit Declaration on Sovereignty in the Arctic'. It reaffirmed Inuit perspectives on the Arctic as a homeland and the specific rights of indigenous peoples. It stressed that, as the world looks increasingly to the Arctic and its resources and as climate change makes access to circumpolar lands easier, the inclusion of Inuit as active partners is central to national and international deliberations on Arctic sovereignty, development and protection.

In the contemporary Arctic, indigenous peoples and the organizations that represent them draw upon traditional cultural values as well as traditional knowledge, but also their ambitions for the future, to define and articulate their interests and claims. They use their knowledge as a political lever to influence policymakers and to empower themselves so that communities can be involved in decision-making processes concerned with the future of natural resource use and environmental protection, as well as claiming the right to determine the course of their economic development. The work of ICC and other Arctic indigenous peoples' organizations often aims to challenge the authority of the state alone as the basis for environmental governance, draws attention to the processes and impacts of development and asks searching questions about the potential benefits. The ICC has taken a lead in pressing regional and national governments to offset the impact of social, economic and environmental change and in persuading governments to work on implementing measures for environmental protection and sustainable development. Notable success stories include the organization's central role in the negotiation of the global Stockholm Convention on the Elimination of Persistent Organic Pollutants. Following the negotiations, the ICC continued to lobby states to ratify the convention in their national legislatures. The convention entered into force in May 2003 and the ICC continues to work to ensure that the convention's obligations are implemented.

In the United Nations, the Permanent Forum on Indigenous Peoples is a body of 16 representatives, half of them nominated by indigenous organizations and half by UN member states, that meets annually to examine indigenous issues. It makes recommendations to the UN Economic and Social Council. Arctic indigenous representatives – particularly Inuit – have played a significant role in the Permanent Forum, demonstrating how global indigenous movements can find ways of negotiating at international levels and ways of representing and intervening in not only the Arctic as a geopolitical region but also in global debates – another way, therefore, that the Arctic can be re-imagined and repositioned in explicitly global terms.

Connected and Contested Polar Spaces

Whatever claims might be made for the relative remoteness or geo-graphical inaccessibility, the Polar Regions are today enmeshed in a series of tensions and contradictions that bring to the fore the uneasy coexistence of political-territorial colonization, resource exploita-tion, scientific research and the maintenance of managerial regimes, both global and regional in scope. Intensifying resource extraction (both living and non-living resources) and competition for access to areas within contested coastal waters and open seas/deep waters does exist (Bert 2012, and more critically, Craciun 2009; Nuttall 2012a). At best, the Arctic and Antarctic might continue to be managed proactively with due regard to existing managerial regimes such as the Commission for the Conservation of Antarctic Marine Living Resources (CCAMLR) and UNCLOS coupled with an appreciation for the so-called 'Lisbon Principles' regarding the sustainable govern-ance of the oceans (Costanza et al. 1998). In the Arctic, traditional indigenous knowledge and hunting and subsistence cultures might also serve as a coda to more industrial-scale exploitation in areas such as fishing and mining.

Mindful of the distinctiveness of the Polar Regions, there is one area where they face a common geopolitical and governance chal-lenge. In both regions, we see a subset of states asserting their rights on the basis of territorial sovereignty. In the Arctic, this stems from established sovereignty over the metropolitan territory of surround-ing states (the so-called Arctic 5), in the Antarctic from the assertion of rights by seven territorial claimants, which are not generally recognized by the international community. In varying forms, the central polar contest is the claim that particular rights adhere to these territorial states versus claims to wider global and, in the case of the Arctic, indigenous, rights. A select group of coastal and claimant states appear eager to assert their territorial-sovereign rights at the apparent expense of other states who wish to preserve open access (or arrangements that preserve *their* access), to shape resource regimes, to protect mobility/transit rights and, where possible, to contain the sovereign rights of coastal/claimant states by ensuring that the freedom of the high seas and transit passage is not compromised.

In the case of the Southern Ocean, the governance challenges are formidable as changing technologies, emerging global markets, ongoing commercial development in areas such as tourism and biological prospecting escalate resource activity. Such an escalation will unfold in a context in which:

> [t]he most immediate conservation threats to species, ecosystems, and resources around the Antarctic margin are consequences of regional warming, ocean acidification, and changes in sea-ice distribution. Marine resource extraction may exacerbate these threats. Current information suggests that toothfish and krill are particularly at risk into the future, but the full extent thereof is unclear due to the lack of comprehensive understanding of their life histories and spatial dynamics and difficulties in obtaining such information. (Chown et al. 2012)

Steven Chown and colleagues pose the question as to whether the Antarctic Treaty System and its constituent legal instruments such as the Protocol on Environmental Protection can cope with such change, especially as by 2048 onwards it will be possible to review the current prohibition on mining. Similar questions might be posed about the Arctic as well, as rooted and routed parties negotiate their way around, through, below, back and beyond the Polar Regions.

— 5 —

NEW RESOURCE FRONTIERS

Over the last ten years or so, academic research, policy documents and media reports have reinforced with increasing frequency the argument that the world is casting its gaze on the Polar Regions as never before. Although some observers were warning about a resource 'rush' in the midst of the 'Second Cold War' (Roucek 1983), as Michael Watts and his colleagues (2014) note, there is no shortage of contemporary 'oil talk' when attention turns to frontier regions, including the Arctic, deep ocean and 'new' offshore regions off the African, South American and Eastern Mediterranean coastlines. We can expand 'oil talk' into a more general 'resource talk' when thinking about how mining technologies, title deeds, engineering companies, infrastructure, finance, insurance, combined with conjecture about resource pricing and future demands, generate new patterns of affluence, effluence and influence. The Polar Regions are caught up in an abundance of 'resource talk'; fish, minerals, whales, timber, and oil and gas.

While Antarctica is in the news when ice shelves break off from ice sheets and Japanese scientific whaling resumes, making headlines about climate change and cetacean protection respectively, the Arctic has come under sustained global scrutiny as a place emerging as the world's 'newest' resource frontier, albeit a scrutiny which seems inattentive to longer histories of mining, sealing and whaling. In an article titled 'Rushing for the Arctic's riches', published in *The New York Times* on 7 December 2013, Michael T. Klare exclaimed that, 'Approximately 13 percent of the world's undiscovered oil deposits and 30 percent of its natural gas reserves are above the Arctic Circle, according to the United States Geological Survey. Eager to tap into this largess, Russia and its Arctic neighbors – Canada, Norway, the United States, Iceland and Denmark (by virtue of its authority

115

over Greenland) – have encouraged energy companies to drill in the region' (Klare 2013). As well as touching on the intensification of oil and gas exploration, like many other commentators on Arctic geopolitical themes who point out how Arctic states posture and assert their claims to control over circumpolar places and spaces, Klare suggested that the risk of conflict over rights to contested and mainly offshore territories is likely to grow. He retraced increasingly well-worn academic and journalistic ground, as many before and since have also done, producing alarmist reports of Arctic meltdown and the inevitable development of mining and hydrocarbon extraction in the wake of diminishing sea ice. Moreover, his focus on the Arctic states and their interests draws attention to offshore Arctic regions and, as a consequence, away from other kinds of schisms that persist within those northern states involving corporations, indigenous peoples and regional governments. At the same time, environmental groups such as Greenpeace are warning that the Arctic needs to be 'saved' from further industrial extraction. Under this vision of the Arctic as resource frontier, there lingers an 'El Dorado' complex in which it is only matter a time before such resources are exploited.

Several months before Klare's piece, Amy Crawford pondered in *Smithsonian Magazine* whether Arctic ice melt would result in 'a 21st-century version of the Great Game, which Russia and Britain played among the mountains and deserts of Central Asia in the 19th century. The prize then was the riches of India; today, it's new shipping routes and untapped natural resources, including an estimated 13 percent of the earth's oil and 30 percent of its natural gas' (Crawford 2013). Around the same time as Klare's article appeared, Canadian broadcaster CBC reported on Canada's claim to an extended continental shelf stretching towards the North Pole in terms of 'Arctic riches' (Paris 2013), while *The Guardian* expressed worries over Greenland's ambitions for 'mineral riches amid fears for a pristine region' (Macalister 2014). Over a year before Klare's piece was published, an article by Elisabeth Rosenthal appeared in *The New York Times* on 18 September 2012 that recycled the trope of the Arctic warming under rapid climate change ushering in a new race for resources in the Far North. Reporting from Nuuk, Rosenthal wrote that 'with Arctic ice melting at record pace, the world's superpowers are increasingly jockeying for political influence and economic position in outposts like this one, previously regarded as barren wastelands' (Rosenthal 2012). She went on to say that 'at stake are the Arctic's abundant supplies of oil, gas and minerals that are, thanks to climate change, becoming newly accessible along with increasingly navigable polar shipping shortcuts'. Much of the article's focus is on China 'becoming a more aggressive player' in the Arctic, 'provoking alarm among Western powers'. It is an example of a trend in journalistic writing, as in more popular books on the Arctic with titles such as *The Arctic Gold Rush*, but also in the scholarly literature, that portrays the Arctic as the last frontier. Along with similar reportage and opinion pieces on struggles for 'hidden' and 'vast' Arctic resources being 'unlocked' and becoming 'accessible' as the sea ice and permafrost melts, these are but a small representative sample of a range of voices contributing to a contemporary narrative about the Arctic that epitomizes a rather crude form of earth-craft and statecraft. It is not inevitable that just because oil, gas and mineral resources are found that it follows they will be commercially exploitable – a point that is often lost on those eager to portray the Arctic as a treasure house writ large. And it is not inevitable that, just because sea ice thins or retreats, the maritime Arctic becomes a more benign operating environment. Community reaction in the Arctic also matters as well as long term trends in markers and pricing (figure 5.1).

What complements 'resource talk' is a form of 'frontier-speak'

Figure 5.1 Protests Against Uranium Mining in Greenland
(Mark Nuttall)

117

where the promise of great resource potential has been a recurring feature of non-indigenous encounters with the Arctic and Antarctic. What makes it interesting is how earlier, contemporary and even future frontiers coexist with one another. Standing on the shoreline of the subantarctic island of South Georgia, for example, one can see the rusting whale oil tanks onshore from previous decades, pick up whale bones that litter the shoreline, and gaze at the ocean and imagine fishing vessels harvesting fish stocks around the island. While it has yet to happen, and may never do so, there has also been specu-lation that beneath those waters there may lie oil and gas resources that a British government, which enjoys sovereignty over the disputed island, may yet exploit. As a counter-claimant of both the Falkland Islands and South Georgia, Argentina in particular dislikes this kind of 'frontier-speak', even if it is just imagined at this stage.

The Arctic and Antarctic as 'resource frontiers' have proven a useful, even necessary, accompaniment to statecraft, business and industry, indigenous communities, environmental campaigners and scientists. While the object, area and importance of the resource fron-tier varies, 'frontier speak' creates the very thing that it names and along with that a bounty of opportunity, challenge and risk for all those around it.

Ghostly Geographies and the New Arctic

Global concern over the impacts of climate change on the Arctic is often tempered by excitement related to the resource extraction pros-pects the region offers. This new Arctic in an era of climate change is imagined as a region of immense potential, with a diverse range of stakeholders, including mining and hydrocarbon businesses and investors and non-Arctic states. This marks a significant representa-tional shift from the 1990s and early 2000s when the agenda set by the AEPS and Arctic Council prioritized issues related to conservation, pollution and climate change – the region's stakeholders were usually seen then as primarily Arctic states, indigenous peoples, scientists and conservation groups. During this time, the Arctic was increasingly considered a region at risk as well as a region of risk, in which conser-vation and the protection of environments and resources, along with sustainable development, were key concerns (Nuttall 1998).

In the second decade of the twenty-first century, however, the items on the circumpolar agenda have gradually become more business-focused. The intense interest in economic development possibilities is evident in the proliferation of conferences and industry events promoted by those eager to transmit the opportunities for commerce and business in a warming Arctic. The likelihood of shipping routes transecting previously inaccessible ice-covered waters and the development of extractive industries in what have until now been relatively remote and under-developed areas is often championed at these events. Initiatives such as the Arctic Economic Council (previously the Circumpolar Business Forum), which was established during Canada's chairmanship of the Arctic Council, suggest that the council is moving in a new direction. The Arctic Economic Council, according to the Arctic Council, reflects the fact that, in the Kiruna Declaration, signed at the eighth Arctic Council Ministerial Meeting in May 2013, ministers from the eight Arctic Council states 'recognize[d] that Arctic economic endeavors are integral to sustainable development for peoples and communities in the region, desire[d] to further enhance the work of the Arctic Council to promote dynamic and sustainable Arctic economies and best practices, and decide[d] to establish a Task Force to facilitate the creation of a circumpolar business forum'. It is intended to provide a 'business perspective' and 'foster business development in the Arctic' (Arctic Economic Council 2014). Some indigenous leaders have welcomed the introduction of the forum.

As Dittmer et al. (2011) point out, an increasingly wide range of actors and institutions are engaged in space-making practices in and about the Arctic today. As an indication, an Arctic Summit series organized by *The Economist* is billed as gazing out to 'new horizons for trade and economic development' at a time when 'the Arctic race is heating up',[1] while a new assembly called the Arctic Circle aims to support, complement and extend the work of the Arctic Council by holding annual meetings and bringing together 'a range of global decision-makers from all sectors, including political and business leaders, indigenous representatives, nongovernmental and environmental representatives, policy and thought leaders, scientists, experts, activists, students and media'.[2] In this 'new Arctic' narrative, the Arctic is on the verge of a transformation into a transnational space (rather than merely circumpolar), firmly embedded in a global economy.

This ignores the fact that, as we discussed in chapter 2, the world has reached into the Arctic for centuries and that globalization is nothing new in the region. While the Arctic is being imagined as a new frontier for oil, gas and mineral extraction, it may be more accurate

---

[1] http://www.economistinsights.com/sustainability-resources/event/arctic-summit-2014
[2] http://www.arcticcircle.org/mission

to see this as a reimagining of the Arctic as a frontier – or perhaps as a re-frontierization process. The articulation and representation of Arctic places by indigenous peoples as indigenous homelands from the 1970s onwards, as intimate worlds of human–animal–environmental relations and intricacies, of being and becoming, did much to counter the prevailing view at the time of the Arctic as empty space, wasteland, wilderness and frontier. Thomas Berger's report from the Mackenzie Valley Pipeline Inquiry, *Northern Frontier, Northern Homeland*, published in 1977, emphasized that the Mackenzie Valley and Mackenzie Delta regions in northern Canada were the social, cultural and spiritual homes of a large number of Aboriginal people, in addition to being areas of increasing economic importance to industry, keen to satisfy energy demand in southern markets. Berger's recommendation to the Canadian government that a pipeline should not be developed until indigenous land claims were negotiated and settled was predicated on the importance of recognizing the North as a homeland rather than a wild frontier ready for industry to develop. Yet, as evident in the contemporary narrative about the new Arctic as a place of possibility and opportunity, the perception of the region as a resource-rich frontier is enduring (Nuttall 2010).

Such geopolitical agitation and speculation does not occur within a historical and geographical vacuum. There have been past moments of 'resource talk' and 'frontier speak' and geographers and sociologists such as Avery Gordon (1997) have alerted us to the way in which the unfinished business of the past intrudes into the present. More specifically, we might understand the past in two distinct senses. First, it returns in some form to make demands on the present, and is thus fundamentally unsettling. Second, it allows us to imagine something that might have been lost or barely present. These understandings of the past can also be profoundly geographical in the sense of where the past might come from and where we might address or confront these ghostly traces. The Arctic as both resource space and energy frontier are two spectral components of the ghostly geographies of the region. As Andrew Stuhl noted in his examination of 'New North' discourses between 1910 and 2010, 'proclamations about the North and associated evaluations of history and geography often licensed political and corporate influence while delimiting the colonial legacies already apparent in the North' (Stuhl 2013: 96). But as Emilie Cameron cautions us,

> What does it mean, then, to be 'haunted' in a decolonizing settler colony like British Columbia [and we might add the Canadian North and the

Arctic more generally]? Who is haunted in these stories, and who or what is doing the haunting? What kind of future might these haunt-ings demand? Do they signal, as Derrida intended, a recognition of the always unfinished and unfinishable in our relation to the present and past and, by extension, a sense of generosity and hospitality towards ghosts? Or do they, as Sarah Ahmed has argued in relation to white guilt in postcolonial Australia, constitute yet another self-referential engagement with the colonial past, in which the experiences and desires of the settler occlude consideration of other desires and possibilities? (Cameron 2008)

Resource Frontiers

The contemporary discourse about a scramble for the poles contin-ues to draw upon and take substance from earlier encounters and understandings of the Polar Regions as resource-rich spaces awaiting exploration, exploitation and development. Initially, however, north-ern regions themselves were not objects of speculation, but places to venture into and cross through en route to other places as destina-tions of trade and commerce. From the late fifteenth century, Arctic exploration originally had as its goal the discovery of northern mari-time routes between the Atlantic and Pacific oceans to the East – the Northwest Passage around North America and the Northeast Passage around Asia – shorter sailing distances for European traders to access the spices, silks and other riches of the Orient.

Robert McClure, under whose command HMS *Investigator* is cred-ited with first charting and traversing part of the Northwest Passage in the 1850s (whilst searching for the lost Franklin expedition), wrote of 'arduous struggle during the course of three hundred years' to find such a maritime passage (McClure 2013 [1856]: 1). Sponsored by King Henry VII of England, Italian navigator John Cabot sailed from Bristol in 1497 to seek a more northerly route to the Indies. His explorations in what is now eastern Canada lay south of the Arctic, and little is known about them, but his voyage was probably the first European encounter with the North American mainland since the Norse sailed from Greenland and reached what they called Vinland in the eleventh century. Following Cabot, the Portuguese sailed farther north and in 1500 Gaspar Corte-Real visited southern Greenland. The first venture to find a Northeast Passage over Eurasia, led by Englishman Sir Hugh Willoughby in 1553–4, ended in disaster when Willoughby and his entire company died, most likely of scurvy, in Lapland. The Company of Merchant Adventurers of England

121

financed three expeditions to the Arctic led by Martin Frobisher in the 1570s in search of a Northwest Passage to the Orient, while John Davis explored parts of the west Greenland coast and the eastern coasts of Canada in the 1580s. English place names on the map of north-west Greenland, such as Sanderson's Hope for the passage just south of Upernavik, named for William Sanderson, a London brewer and one of the expedition's sponsors, mark Davis's progress and anticipation, just as they commemorate the ambitions of those who voyaged after him.

Frobisher caused a sensation when he returned to Elizabethan England in 1576 from his first voyage. Not only did he believe he had found an entrance to the Northwest Passage, he brought with him a cargo of black rock from Baffin Island which assayers declared was laden with gold. Frobisher's second and third expeditions involved his crews in mining operations and they returned to England with over 1,000 tons of ore. The assayers were mistaken, but stories of vast mineral wealth, of mountains of silver and gold in Greenland and in Baffin Island and other Arctic islands were to abound in Europe for the next couple of centuries – Denmark's King Christian IV notably sent expeditions to Greenland in search of silver in the early 1600s – until organized mineral exploration hinted at the realities and possibilities of real discoveries.

Many regions and people throughout the circumpolar world have a history of experiencing the economic, environmental and social impacts of extractive industries (and, as we pointed out in the previous chapter, these impacts often manifest themselves as toxic legacies). Indeed, mining is not a new industry for some northern regions, but has been the single or dominant one in places such as parts of northern Sweden and in Svalbard. And an industry where money has been made and money has been lost. Norwegian and Russian mining operators continue to exploit the coal seams of Svalbard but now struggle to generate much profit. The extraction of cryolite began at Ivittuut in south Greenland in 1854 and a coal mine at Qullissat on west Greenland's Disko Island was opened during the early years of the twentieth century (although the coal was mined primarily to fuel the power stations of the larger west coast towns rather than for export). In North America, the gold rush opened up Alaska and Canada's Yukon in the 1890s, bringing thousands of prospectors from all over the world hopeful of making their fortunes; coal mining in Norway's Svalbard archipelago dates from the early twentieth century; oil production began at Norman Wells in Canada's Northwest Territories in the 1920s and Canada's High Arctic saw oil and gas projects in

122

the 1960s. A massive industrial transformation of the Russian North and Siberia began in the 1920s with the exploitation and movement of timber, coal and minerals to be processed in new northern industrial towns. Today, coal, nickel and copper are mined near Norilsk, and gold, diamonds, uranium and other minerals have been found in great quantity in other northern parts of Russia – for example cobalt and copper in the Pechenga fields along the Finnish border. High grade iron ore has been mined at Kiruna in northern Sweden and copper and gold have been mined in south Greenland, while in sub-arctic Canada and Alaska there is a wealth of mineral resources – for example uranium and diamonds in the Northwest Territories and zinc and lead in north-west Alaska.

Most descriptions of resource frontiers, says Anna Tsing, 'take the existence of resources for granted' (Tsing 2003: 5100). Edward Barbier (2011) argues that a frontier is an area or source of unusually abundant natural resources and land that has to be seen relative to labour and capital and it is the abundance or relative scarcity of natural resources, not their absolute physical availability, that is of importance to economic development. The process of frontier expansion, which for Barbier is the exploitation and conversion of resources for purposes of production, has 'been pivotal to economic development for most of global history' (Barbier 2011: 8), but it is the expectation of riches and the promise of the frontier that holds the allure of expansion into areas perceived as wild and peripheral, yet resource abundant. What we are witnessing in the Arctic and other supposedly frontier regions is a process of resource expansion that is increasingly looking in more explicitly volumetric ways. As Findlay and Lundahl put it, the frontier extends vertically downwards into the subsurface rather than something that is 'horizontally extensive as in the case of land and agriculture' (Findlay and Lundahl 1999: 26).

The Arctic as a Hydrocarbon Province

While mining is set to increase in importance in the Arctic, global attention has focused on the North as a hydrocarbon frontier. This requires us to think about the particular and specific qualities of natural resources, in terms of both 'the biophysical particularities and political and discursive peculiarities of oil' (Watts 2001: 191). Despite academic and policy discourses on security and defence, an increasing number of military exercises, and warnings of possible conflict in northern regions, the Arctic is nonetheless considered a geopolitically stable region. It is thus attractive to a global industry, at precisely the moment when the availability of oil, gas and minerals becomes unreliable either because of geopolitical tension and conflict in some parts of the world and/or as a result of resource depletion. Emerging threats to international security arise partly because of this unreliability, demographic growth, industrialization of the developing world, and increased competition for energy and food resources (Krishna-Hensel 2012). As Watts points out, 'oil is inseparable from the largest forms of transnational capital; and not least oil has its own mythos rendered through its fantastic wealth, its all-encompassing power, and as the lifeline to hydrocarbon civilization' (Watts 2001).

When international media reported on the 2007 Russian flag planting expedition, journalists were swift to connect this moment to a broader possibility of the offshore regions of the Arctic being subject to intensive exploration for oil and gas. As we have already pointed out, in the Antarctic, mineral resource activity is banned under the 1991 Madrid Protocol (which entered into force in 1998) and, historically and in more recent times, attention has been focused in the main on living resource exploitation including whaling and fishing. While global attention may not yet be focusing on the potential of Antarctic hydrocarbons (although there have been moments when this was the subject of speculation in the 1970s and 1980s), it is the Arctic and subarctic regions that are considered to be some of the world's last energy frontiers. The United States Geological Survey estimate that 25 per cent or more of the world's remaining untapped reserves could be found in the Arctic heightened the sense of anticipation and indeed expectation about resource extraction from northern lands and waters. But as we have already noted, oil and gas development is nothing new in the Arctic. Most activity so far has involved oil production onshore along the North Slope of Alaska (oil has been flowing down the trans-Alaska pipeline from Prudhoe Bay to Valdez since 1977), in western Siberia and offshore in the Barents and Beaufort Seas. The Alaskan North Slope, Canada's Mackenzie Delta, Russia's Yamal Peninsula and their adjacent offshore areas hold natural gas deposits forecast for development during the next decade or so. Furthermore, oil exploration continues off west and east Greenland: and, as we have also noted, the USGS estimates that Greenland's west coast waters could contain about 110 billion barrels of oil, with at least 31 billion barrels of oil, gas and natural gas liquids in the East Greenland Rift Basins Province, have generated intense interest (Bird et al. 2008). In northern Alaska some oil companies are

124

considering the prospects of exploiting heavy oil with the consistency of thick molasses, which would extend the lifetime of the industry in this part of the state.

But as exploration intensifies and oil and gas development occurs in more of the Arctic, indigenous peoples and other northern residents are increasingly anxious about the growing interest and influence of industry, national governments and the far-reaching impacts of global markets and resource demands on their livelihoods and environments. Many communities are under increasing pressure from resource exploration projects, to sign on to development projects, to negotiate with industry and governments, and to find ways of adapting to a changing environment (as well as new socio-economic circumstances) resulting from the activities of extractive industries. Public participation in large-scale development projects is seen as a necessary prerequisite for the success of a project, for minimizing environmental risk and social impacts, and for strategies of sustainable development. Article 26 of the United Nations Declaration on the Rights of Indigenous Peoples, which was adopted by the UN General Assembly on 13 September 2007, states that 'Indigenous peoples have the right to the lands, territories and resources which they have traditionally owned, occupied or otherwise used or acquired.' Article 32 further asserts that 'States shall consult and cooperate in good faith with the indigenous peoples concerned through their own representative institutions in order to obtain their free and informed consent prior to the approval of any project affecting their lands or territories and other resources, particularly in connection with the development, utilization or exploitation of mineral, water or other resources.' Yet participatory and consultation processes are uneven across the Arctic and even when public involvement is considered a central element of regulatory frameworks, it does not always necessarily work to the benefit of northern residents, something that would have not surprised earlier generations of northern communities (e.g. McCreary and Milligan 2014).

Indigenous Peoples and Energy Development

The Arctic has a history of sensitivity and vulnerability to change, and to the impacts of industrial development, or what Michael Watts might see as an ecological violence perpetrated upon the biophysical world (Watts 2001). It is a place with a fragile ecology where environmental scars from resource extraction take decades to heal, if at all. In addition to direct effects and impacts on vegetation and hydrology, oil and gas development activities have many cumulative effects on traditional resource use practices and local economies. Research has shown that as energy and military development in the Arctic expanded in the second half of the twentieth century, transportation infrastructure (roads, pipelines, airstrips, ports) contributed significantly to surface disturbance and habitat destruction. Pipeline construction, which creates the necessity for roads and thereby leads to easier access to formerly isolated regions, also opens up larger areas for additional resource development. Contrasted to this, as a synthesis of research carried out for the Millennium Ecosystem Assessment argues, between 1900 and 1950, less than five per cent of the Arctic was affected by infrastructure development. Some scenarios suggest that by 2050, 50–80 per cent of the Arctic will be disturbed by development, although this level of environmental impact may occur by 2020 in parts of northern Fennoscandia, and some areas of Russia as large-scale extraction projects either intensify and/or are initiated (Chapin et al. 2005).

In the Russian North, pollution from oil and gas activities affect reindeer pasture and marine and freshwater environments. The construction of pipelines and the actual production of oil and gas also impede access to traditional hunting and herding areas. As Stammler and Forbes (2006) show, industrial development in the Soviet Union caused tremendous damage to indigenous livelihoods. A lack of ecological consciousness at the time led to degradation and pollution of reindeer pastures and hunting and fishing grounds, as well as community displacement that still affect local land use. In the immediate aftermath of the collapse of the Soviet Union, the extent of environmental damage increased, while preventive measures or repairs of facilities were delayed or not undertaken. Widespread ecological damage in Yamal Peninsula in western Siberia is an example of what can happen in an intensively developed oil and gas region, and one where the rights of indigenous peoples are considered to be secondary to the national security objectives of governmental elites who imagine the Russian state to be an 'energy superpower'.

Oil and gas development in Arctic regions may also influence marine mammals. Noise from offshore oil exploration in the Beaufort Sea disturbs bowhead whales and could deflect them from migration routes, making them less accessible to Iñupiat hunters. Oil spills from marine transportation or offshore oil platforms have the potential for widespread ecological damage, particularly in ice-covered Arctic waters. When oil spills occur, the effects may be long lasting. As Gill and Picou (1997) have shown in southern Alaska, the subsistence lifestyles of Native Alaskans predisposed them to impacts of the *Exxon Valdez* oil spill in 1989 and the subsequent clean-up activities. The direct effects included emotional distress and disruption, threats to subsistence activity and consumption because of fears over contamination of marine living resources, and the disruption of harvesting because local people were also involved in clean-up activities. Gill and Picou also showed how the clean-up also directly affected the cultural complex of subsistence, including an influx of outsiders into Native communities, destruction of historical/archaeological sites, racism, disrupted family activities, psychological stress and substance abuse. The long-term legacies of the spill and clean-up continued for several years, including decreased consumption of customary foods and decreased harvesting.

Risks associated with hydrocarbon exploration and production increase with many factors, including water depth, sea ice, icebergs and storms. Spills from pipelines in temperate-zone oil basins in the headwaters of Arctic rivers such as the Ob, Pechora and Mackenzie could also contaminate Arctic waters. Furthermore, black carbon particles and methane emitted from flaring, incomplete combustion in energy production and ship transport (and other industrial processes) result in surface warming when emitted to the atmosphere, and in the case of black carbon when deposited in snow and ice where it further amplifies melting (Sharma et al. 2013).

Oil and gas development activities in the Arctic are of critical importance to the present and future social, cultural, economic and ecological circumstances of Arctic indigenous peoples, and they have both negative and positive impacts and consequences. Concerns arise because of fears of drastic and long-lasting social, economic and environmental impacts, but there are other anxieties because of disputes about the ownership, use, management and conservation of traditional lands and resources in the homelands of indigenous peoples. These issues are at the heart of the Inuit Circumpolar Council's 'Circumpolar Inuit Declaration on Sovereignty in the Arctic', but they are also emphasized in other statements by indigenous peoples who reiterate the need for industrial developers to recognize their responsibility to indigenous and local communities and to the environment.

Not all indigenous people see energy development in their traditional territories as negative, however, although there are certainly strong and emotional concerns expressed by people who are faced with oil and gas development projects. Iñupiat whaling captains and Greenlandic fishers worry over the presence of seismic survey vessels near whale migration routes and feeding grounds and good fishing areas, while hunters and trappers in the Mackenzie Delta and in boreal forest communities are anxious about seismic cuts and pipelines disrupting traplines and traditional hunting lands, as well as the desecration of sacred sites. However, in Alaska, Canada and Greenland, the settlement of land claims and the introduction of self-government has meant that some indigenous communities have entered into resource development projects through joint ventures with industry and government, impact benefit agreements, and environmental monitoring projects. Although traditional and customary practices remain important to the daily lives of indigenous peoples, oil, gas and mining activities increasingly provide employment in some communities, especially in Canada and Alaska, and are expected to do so in Greenland and Russia. Far from being mere victims of the impacts of industrial development, indigenous peoples are also participants in, and increasingly beneficiaries of, the development of the Arctic resource frontier. In the early 2000s, for example, the Mackenzie Gas Project was promoted as a major energy development initiative that would benefit northern Canadian indigenous communities because one of the proponents of the project, along with Imperial Oil and other major players, was the Aboriginal Pipeline Group (APG). Heralded as a new form of business venture for the indigenous North, the APG came on board as a one-third shareholder in the project and argued that the profits would flow to local communities. Following a lengthy review process and extensive period of public hearings (Nuttall 2010), the project was given regulatory and government approval but has recently stalled due to rising project costs, a decline in natural gas prices, and the development of shale gas in the United States. The Gwich'in Tribal Council, as part of the APG, is just one of many regional organizations in the Northwest Territories hoping that rising prices and an increasing demand for natural gas will revive the project. In some cases, indigenous communities initiate the development of activities related to extractive industries. This is illustrated by the business interests of Alaska Native corporations on the North Slope, by the involvement of Canadian Aboriginal people in the Mackenzie Gas Project, and by the agreements Greenland's Self-Rule government has entered into with international oil, gas and mining companies as a precondition for establishing the economic basis for Greenland under Self-Rule and eventual independence.

Oil and Mineral Rights

Because the political landscape has changed in many parts of the Arctic with the signing of comprehensive land claims and other settlements, responsibility is placed on energy companies to negotiate with indigenous communities and include them in decision-making processes and in environmental and social impact assessments. This may not immediately ease fears of the high environmental and social costs that oil and gas development often leaves in its wake, but it may provide a context for discussion of appropriate strategies for sustainability and environmental protection. It also complicates our understandings of resource frontiers, and the Arctic experience, in all its diversity, stands in contrast to an Antarctic one where the absence of indigenous human communities has made it even easier to imagine the southern polar continent and surrounding ocean as a place of actual and potential subsurface resource abundance.

The question of who has rights over subsurface resources and to their development is often greatly contested. Fondahl and Sirina (2006) discuss how, for much of the late twentieth century, hydrocarbon development in Russia took place on native lands, largely in the western Siberian oil fields on Khanty and Mansi traditional territories and the north-western gas fields on Nenets homelands. More recently, in the far east of Russia, controversies over the exploitation of Sakhalin Island's oil reserves have challenged the territorial rights of the Nivkhi, Evenki and Uilta. While Russia's indigenous peoples have received some benefits from oil and gas development, they do not have rights to subsurface resources, and generally the costs and the negative impacts of development on society, culture, environment and wildlife outweigh whatever positive experiences can be identified and chronicled. Land rights, then, remain at the centre of concern in northern Russia. Unlike in the Alaskan or Canadian North, all land in the Russian North is owned by the state, and indigenous land rights as such exist only to a rather limited extent. However, as Stammler and Forbes (2006) explain, since 1999, as a result of indigenous activism and involvement of Russian anthropologists, three federal laws have been passed that concern indigenous minorities. One guarantees the rights of indigenous minorities, the second stipulates the establishment and tasks of indigenous community enterprises, and the third deals with what are defined as 'territories of traditional nature use'. Theoretically, these laws provide the basis for the continuation of indigenous economic activity in the North protected from industrial development. However, processes for operationalizing them remain fraught with difficulty, as regional and municipal administrations have to issue legal acts to stipulate the details of their implementation. In the North American Arctic, the implementation of land claims in Alaska, and in Canada's northern territories of Yukon, the Northwest Territories and Nunavut, has provided an institutional framework for mitigation and compensation, as well as for the involvement of indigenous peoples in oil, gas and mining activities. Furthermore, comprehensive land claims in northern Canada, such as the Inuvialuit Final Agreement (1984), the Gwich'in Comprehensive Agreement (1992) and the Sahtu Dene and Métis Agreement (1994) have given indigenous people subsurface mineral rights to specific areas of land. In the Northwest Territories, for example, extractive industries such as diamond mining and oil and gas exploration have also provided substantial cash infusions to communities in some cases. In Russia, the benefits which Northern Autonomous Districts receive from having oil and gas companies on their territories does mean taxes from oil revenues enter the region and constitute a major source of revenue. While current legislation requires companies to compensate local indigenous communities on whose lands they operate, in reality, this often means individual arrangements are worked out which result in payments and goods delivered directly to the indigenous families working with reindeer or dependent on hunting and fishing rather than to the wider community or group. Murashko (2008), however, points to the lack of public participation in regulatory processes concerned with large-scale development in Russia, while Fondahl and Sirina (2006: 65) have also described the frustration expressed by indigenous people because they were unable to express their views and to have a voice in oil development projects. Discussing an oil pipeline project in eastern Siberia, they point out that 'Many are not categorically opposed to the project but rather want to ensure that ecological safeguards are in place, and that they benefit from the construction of such a project through their homelands, whether through compensation payments or through employment opportunities.'

Despite such situations, it is not a simple issue of traditional cultures facing the onslaught of change and disruption brought by industry. In some parts of the world, indigenous business and community leaders see and reap benefits from such development and, indeed, companies owned by indigenous people are involved in the energy sector. This is particularly the case in Alaska and northern Canada, where subsurface rights have been negotiated and community and regional corporations have been created. As Greenland acquires a greater degree of political autonomy, government and business elites are hopeful that oil, gas and mining will become major drivers of the economy, with some political leaders expressing ambitions that resource extraction will generate revenues that will help Greenland achieve full independence from Denmark (Nuttall 2013). The international resources community has identified the potential for Greenland to be a significant source of new mineral and petroleum development, with the opening of new mines and heightened interest in exploration opportunities offshore Greenland in recent years. In 2008, the Danish–Greenlandic Self-Rule Commission concluded a series of negotiations on mineral rights, ownership of subsoil resources, and the administration of the revenues from mining and hydrocarbon development. The commission emphasized that minerals in Greenland's subsoil belong to Greenland and that the country has a right to their extraction. Under the new political arrangement of Self-Rule, which was instituted on 21 June 2009, the Government of Greenland has been granted the rights to administer revenues from the energy and other extractive industries, which illustrates that the relationship between indigenous peoples, governments and the mining sector is not straightforward.

Accelerating the Greenlandic Resource Frontier

Today, like many other parts of the North, Greenland is being represented and positioned, but also marketed, as a major new frontier for oil, gas and mining. Climate change is often cited as contributing to the conditions that make resource development possible as accessibility to places where minerals and hydrocarbons are located becomes easier, and there is increasing optimism among Greenlandic elites about the prospects of Greenland becoming a major global supplier of raw materials, including rare earth elements. The development of oil, gas and mineral resources has become a stated aim of the Greenlandic government and the exploration for, and exploitation of, non-renewable resources has been a cornerstone of government policy for several years. Over the last few years, foreign companies have been setting up offices in Nuuk, the country's capital, making it a place bustling with geologists, prospectors, consultants and others as it transitions into a base supporting extractive industries in administration, logistics and exploration, while other towns along the west coast experience the movement of foreign workers transiting to and from oil exploration drilling rigs (Nuttall 2012a, 2015). To take one example of the nature of this interest, London Mining, a British-based company that developed mines for the steel industry, had proposed to develop an iron ore mine (with Chinese financial backing) at Isukasia, deep into the fjord system some 150 km northeast of Nuuk. Whether the Isukasia project will eventually go ahead is another matter, as a fall in commodity prices and dramatic drop in the company's shares – complicated by the impact of the Ebola epidemic in West Africa on its only global operation, an iron mine in Sierra Leone – led to London Mining going into administration in October 2014. Chinese mining company General Nice took over the Greenland project in January 2015, but regardless of the project being realized in the near term – an iron mine at Isukasia has been speculated on and dreamt about by several interested parties since the 1960s – London Minings's activities illustrate some general issues that persist in Greenland about how a future society and economy are to be imagined in relation to extractive industries. Concerns over development plans for Isukasia that have been expressed by some local people and organizations have centred on the social and environmental impacts (there are worries a mine and its associated facilities and infrastructure, including a road, a pipeline and a harbour, would disturb sensitive ecosystems and reindeer habitat, as well as exclude hunters and fishers from access to an area that would effectively be marked off as an industrial zone), but they also largely settle around discussion about an absence of community consultation and public participation in decision-making processes as well as inadequate social and environmental impact assessments that is not only particular to the Isukasia project (Nuttall 2013). To take another example, in 2007 the Greenland government drew up a memorandum of understanding with the American aluminium company Alcoa. This opened the door to Alcoa moving through initial planning and assessment phases to build an aluminium smelter (with two associated hydroelectric dams) near the town of Maniitsoq on the west coast with an annual production capacity of 340,000 tonnes of aluminium ingots. The government established a company, Greenland Development (which has since been dissolved), to assist in the various activities related to the project, and the limited public debate in Greenland has so far focused on the environmental and social impacts. Of particular concern to local opponents, and something that has been debated vigorously in the national media, is the potential influx of migrant workers to build the dams and the smelter, as well as large mining projects such as the Isukasia venture and other mines in the Nuuk Fjord and elsewhere along Greenland's coasts. In response, people are saying they want to have more information about large-scale projects and to be able to have a stronger voice in discussion about Greenland's future. While the prospects for hydrocarbon development remain unknown, it is mining that, in the words of many politicians, business leaders and contractors, 'must secure Greenland's future as a modern industrial nation'. In anticipation of this, a new harbour is being constructed in Nuuk to allow for increased capacity for container ships and for international companies and plans are on the drawing board for a new airport.

Around the time when Greenland's Bureau of Minerals and Petroleum (BMP) took over control of the minerals industry in 1998, there was a decline in the number of exploration licences issued, as well as a general decline in the interest international mining companies were showing in Greenland and its mineral resources – due largely to poor commodity prices and the high costs associated with exploration. However, international companies are being encouraged by the BMP and other government institutions to invest in Greenland and to explore for and develop non-renewable resources. During the last few years there has been a significant rise in the number of mineral exploration and mining development licences that have been granted, even one named 'Lady Franklin' to the west of Nuuk (figure 5.2). Exploratory work for oil has taken place off west Greenland and seismic activities have expanded to Baffin Bay and the north-west coast, in areas that are close to small hunting and fishing communities. Controversially, exploration is also slated to take place off north-east Greenland, in waters adjacent to the world's largest national park. People in many parts of Greenland, from large towns to more remote areas, are noticing the seasonal arrival of prospecting crews, seismic survey vessels and migrant workers. Every year sees increased seismic survey activities in particular, provoking concern from fishers, hunters, biologists and environmental NGOs, as well as from emergent grassroots organizations, about the impacts on the environment and on marine and terrestrial animals.

While energy companies and Greenlandic politicians remain optimistic that it is only a matter of time before discoveries of commercially viable oil are made, mining activities, energy development plans and Alcoa's ambition to build an aluminium smelter and hydroelectric dams on the west coast are the projects that are the closest to being realized. Political and social debates within Greenland about the nature of such development have become dominant in daily discourse. Public disquiet over lack of appropriate consultation (and criticism over the absence of information about planned megaprojects) has led to forms of resistance and protest in which demands

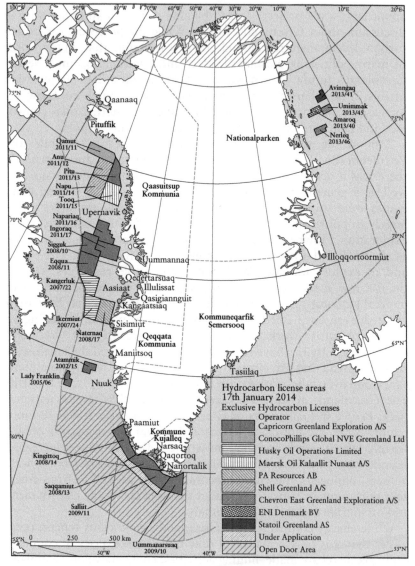

Figure 5.2 Map of Hydrocarbon Licensing in Greenland
(Government of Greenland)

for legitimate public engagement in democratic and transparent discussion and debate over extractive industries are expressed, and this challenges both the representations and governance of resource development.

134

The issue of public participation and consultation processes for large-scale projects remains one that divides communities, government and industry and dominates discussion about how Greenland should not only prepare for extractive industries, but how society will govern resource development in the future. As protests and resistance in relation to possible uranium mining in south Greenland show, there is a widespread sense of mistrust of, and low public confidence in, government and in public bodies, and pro-development discourse nurtures feelings that express scepticism about the real interests of resource stakeholders and multinational companies seeking to drill oil wells and dig mines (Nuttall 2013). At the same time, however, while self-government and a healthy financial return from resource development would mean a greater degree of independence from Denmark could be possible, the construction and operational phases of large-scale projects may also mean the beginning of new forms of dependency relations as multinational corporations broker deals and increasingly exert influence over Greenlandic politics and business (Nuttall 2012a, 2013).

Critically, the making and acceleration of the Greenlandic resource frontier means the classification of certain landscapes and waters as 'zones of sacrifice', to be set aside for extensive development and intensive resource exploitation. In this process, the specificities of place, the histories of use, travel and movement, and the complexities of human–environment relations are not just ignored but erased. In the Nuuk Fjord area, for instance, place names and local narrative accounts of seal hunting, whaling, tracking reindeer, fishing or travelling by boat, walking across the land, plant and berry picking, and camping attest to the historical and contemporary use of the area, and inform multi-spatial thinking and anticipation about its future use. The area is a complex assemblage of past and present cultural landscapes and environments. When one travels through and around it, one becomes aware of multiple layers and multiple meanings in it, of the confluences of interior landscapes, inlets, glaciers, rivers, wider stretches of open water, local knowledge, place names, stories and narratives, history and biography, and geology and society. But such rich intergenerational connectedness with the area contradicts and challenges how Nuuk Fjord is imagined, described and represented by industry consultants and politicians ambitious for the development of mines, and for whom the area becomes a wilderness, empty of human presence. Local experiences and knowledge are erased by the production of technical knowledge and in political and industry discourses about Greenlandic environments and subsurface resources

(Nuttall 2012c, 2015). Isukasia and places like it, which promise the extraction of riches from the earth, become central to Greenlandic ambitions for state formation and possible independence. As sacrificial zones, however, such places in which people dwell, move in and around, and which nourish them, must first be rendered empty and made wild (Nuttall 2013, 2015), their long histories of use and the tangled and intersecting webs of meaning ignored in a process of reconstruction into vast, expansive, deterritorialized, open spaces (e.g. Kuletz 1998).

Deferring or Delaying the Antarctic Frontier?

If there is acceleration in some parts of the Polar Regions, there is evidence of deferral in other areas, namely Antarctica and the Southern Ocean. This is not to claim, however, that resource exploration and exploitation have not been strong and recurring features of the human encounter with the Far South. Exploration, discovery and exploitation were bedfellows in the nineteenth century as sealing and later whaling became multinational enterprises, serving international market demand for seal pelts, whale oil and the like. British, American, Chilean, Argentine and Norwegian operators played their part in supporting and sustaining these resource-intensive activities, and in catalysing the British in particular to make the first territorial claim to the Antarctic in 1908. Regional islands such as the Falkland Islands and South Georgia were essential accomplices to the exploitative trade in seals and whales. While later developments such as pelagic whaling changed the administrative geographies of British imperial control (in other words diminishing the importance of strategic gateways in controlling the trade as much was now occurring in the high seas), it is worth bearing in mind that commercial whaling was still a feature of Antarctic activity until the 1960s.

The Antarctic and subantarctic shorelines and hinterlands bear witness to this resource frontier-like environment. The sealing and whaling industries were rapacious in their outlook and practice. Hundreds of thousands of seals and whales were captured, killed, processed and exported to markets in North America, Europe and China. Evidence of settlements in South Georgia and along the Antarctic Peninsula still remain to this day, as do the ghostly remains of those processing plants. As commercial whaling was diminishing in importance in the post-1945 period, another resource frontier was being imagined. This time the source of conjecture and speculation involved non-living resources. While the Antarctic was long thought to contain minerals and resources such as coal, there was mounting interest in other objects such as uranium, zinc and iron ore. The colonization, claiming and contestation of Antarctica and the Southern Ocean were fuelled by resource speculation. By the 1950s, it was common to read that the Antarctic was a treasure house, waiting to be discovered and exploited. With the mounting interest of the US and Soviet Union as potential claimants, the negotiators responsible for the drafting and implementation of the 1959 Antarctic Treaty deliberately excluded minerals and mining rights from its provisions, echoing a decision by the IGY Antarctic parties to avoid geological research, which might be construed as mineral resource evaluation. In order to secure consent, territorial claims and any associations with ownership and mineral exploitation had to be deferred, in this case by ignoring it. Thirty years later, such a position was not tenable, however.

As the 12 signatories deferred a mineral resource frontier at least in 1959, so another resource frontier emerged from within the Southern Ocean. Fishing fleets increasingly turned to more southerly waters as existing fishing grounds were more extensively exploited. Fish and other living organisms such as krill were subject to ever greater exploitation and the ATS addressed living resource management via a convention negotiated in the late 1970s, which was intended to balance conservation and exploitation on the basis of guidance from fisheries science. In the midst of fisheries negotiations, attention also turned to the oil and gas potential of the Ross Sea. 'Frontier speak' turned to another kind of resource. A USGS report in 1974, on the basis of some preliminary seismic research conducted a year earlier, estimated that there might be deposits of 45 billion barrels of oil on the continental shelves in West Antarctica. The US Energy Information Administration in 1975 followed up with an estimate of 50 billion barrels of oil in the Ross and Weddell Seas. All of which, while unquestionably a form of resource guesswork, repositioned the Antarctic as increasingly a resource frontier space and not, as in the case of the late 1950s, a peaceful and rule-based scientific laboratory.

The Antarctic and the Southern Ocean represented as multiple frontier spaces, with a series of overlapping resources, created a series of long-term consequences. The region became ever more internationalized, as newer countries joined the ATS and other countries, notably Malaysia, used the UN General Assembly to raise questions about how the Antarctic was governed. Both living and non-living resources of the region attracted attention – a living resource

137

management convention entered into force in 1982 (CCAMLR) and for six years the ATS attempted to negotiate a mineral resources convention (CRAMRA) on the basis that it was better to have a rules-based system in place before mining occurred in Antarctica. The net result of those negotiations, however, was to encourage Third World states and environmental groups to question not only the governance arrangements but also 'frontier speak' itself.

In retrospect, the 1970s and 1980s were a period when the resources of the Antarctic and Southern Ocean were being viewed in three different ways. First, as something that could be simply governed by those who were party to the ATS. Second, as something that needed the involvement of the international community. Finally, the wilderness and aesthetic qualities of the region meant that there was a pressing need to block resource exploitation – to stop it from turning the Antarctic into a resource frontier of the sort seen in Africa and Asia. In 1991, bowing in part to opposition within the ATS and wider public criticism of the more exploitative forms of 'frontier speak', the Madrid Protocol was negotiated (entered into force in 1998) and Article 7 prohibits all form of mining and mineral exploitation. A review of the protocol was possible, and there is scope for a review conference some 50 years after it enters into legal force (i.e. 2048).

While mining for now is not on the table politically, resource extraction continues nonetheless. Southern Ocean fishing is big business and there is a lucrative trade in fish such as the Patagonian Toothfish. Despite increased investment in maritime surveillance, fisheries conservation measures and market controls, illegal, unregulated and unreported (IUU) fishing continues to trouble coastal states such as Argentina, Australia, New Zealand, the UK and South Africa in the coastal waters around their islands in the Southern Ocean. Countries such as Australia have invested considerably in maritime patrols and policing in the last two decades, as part of a determined effort to protect their 'southern' frontier from illegal fishing vessels and Japanese whalers. Attempts within CCAMLR to establish marine protected areas (MPAs) in the Ross Sea have struggled to achieve consensus agreement in large part because some of the parties most implicated in IUU fishing are also members of CCAMLR (e.g. Russia and Korea). What this reveals is that the Southern Ocean is a deeply contested resource frontier with claimant/coastal states eager to promote stewardship as a way of shoring up their sovereign interests, while extraterritorial states and their distant fishing fleets are mindful of their rights/interests to fish in international waters.

138

Regulating fishing (and even 'scientific whaling') remains a work in progress, and one which reveals divisions over how to regulate the Southern Ocean with its patchwork of islands, and accompanying coastal states intent on protecting their exclusive economic zones from IUU fishing. Beyond that, however, there is a lack of consensus as to how the Antarctic and Southern Ocean should be conceptualized as resource space and/or conservation space. CCAMLR seeks to reconcile those conceptions through the use of so-called conservation measures but some states complain that conservation is being used to restrict further any sort of reasonable exploitation. Moreover, Russia, China and Ukraine have also published statements intimating that they are interested in exploring and evaluating parts of the polar continent for mineral resource potential. While the Madrid Protocol places restrictions on mineral exploitation, these states have questioned such a prohibition and raised questions about whether there might not be a review process post-2048.

Deferring and delaying the Antarctic resource frontier is only one aspect of course. While the Antarctic has not experienced the kind of mining history emblematic of the Arctic region, it has been imagined over and over again as a bonanza region. And it is not just oil, gas, uranium and coal that might be at stake. As Elizabeth Leane (2012) reminds us, there has been pulp fiction aplenty speculating about harnessing polar ice for commercial gain, discovering and exploiting rare minerals, and utilizing secret geothermal power sources that lurk underneath the polar ice cap. The closest we have at present is the growing industry of biological prospecting, which seeks to take elements of Antarctic life (e.g. Antarctic algae) and transform it into something of commercial value raising troubling issues for the ATS about how scientific and commercial knowledge about Antarctica and the Southern Ocean is exchanged or not. One possibility is that the Antarctic becomes a more secretive resource frontier, a place where remoteness and commercial sensitivities combine well to evade others who seek to monitor, to survey and ultimately to manage.

Conclusion: A Rush for Resources?

The increasing global interest in the energy potential of the Arctic, particularly the resource-rich continental shelf, raises the prospect of a future characterized by disputes over Arctic sovereignty and ownership of territory and resource rights. Much of the projected oil and gas development in northern Alaska and northern Canada, for instance, will take place to satisfy domestic market demand, but it will also be driven by security concerns and be critical for those nations' ambitions for energy independence (Fjellheim and Henriksen 2006). China is investing in Alberta's oil sands industry and is eyeing oil exploration leases off west Greenland as well as seeking to fund Greenlandic mining projects, while the European Union has edged towards an Arctic policy, illustrating how many countries outside the Arctic are looking to the region to meet their future energy needs. European countries are increasingly dependent on Russian energy resources and this influences political and economic strategies and international relations. As the world looks northwards for its oil and gas, territorial challenges are provoking nations like Russia and Canada to reassert their claims over their northern hinterlands. In March 2009, a Russian policy document declared that the Arctic must become Russia's top strategic resource base by 2020. The announcement came after Canada announced its intention to increase its naval presence in its Arctic. Both countries have emphasized that their political and economic futures lie in the Arctic. The Northwest Passage, however, as we noted in chapter 1, is not recognized as Canadian internal waters by the United States and the wider international community which, instead, argues it is an international strait. Russia faces similar issues over the future use of the Northern Sea Route and its right as a coastal state to manage international maritime traffic subject to the provisions of Article 234 of UNCLOS. Once part of a major Arctic transportation system, Russian shipping in the Northern Sea Route has declined, but several countries including the US, EU member states and Japan consider it a potentially vital transcontinental shipping lane between Europe and the Asia-Pacific region.

But as we have also shown, there has always been an element of fantasy to the imaginative geographies of the Arctic and Antarctic as resource frontiers. Both places have been imagined as 'El Dorado'-like spaces, simply awaiting capital, infrastructure knowledge and labour to combine to create resource assemblages capable of mining and moving resources including fish, seals, whales, oil, gas, diamonds and uranium. But the process of transforming the Arctic and Antarctic into resource spaces has never been straightforward as the material economies of production, transformation and consumption produce their own challenges and contradictions. Weather, ice and snow, distance and remoteness play as much a part in this as do global markets, commodity prices, infrastructural provision, political and economic incentives, indigenous land claims and trade campaigning. Some resource frontiers were more intensive than others, and every resource in question produced its own particular configuration of sites, agents and capital.

There never has been one polar resource frontier per se, and sometimes the Polar Regions have been exploited in other ways beyond that of minerals and living resources such as fish. One of the most intriguing developments has been the use of Alaska as a drone-testing frontier. The University of Alaska Fairbanks is one of six official test sites. Already used for wildlife management, search and rescue exercises and military surveillance, drone technology is of interest to native Alaskan corporations such as ASRC Federal who think such unmanned aerial vehicles could aid and abet oil companies to supply Alaskan oil rig installations. It might also offer benefits in terms of servicing goods and supplies to more remote Arctic communities, and as such further facilitate energy projects and wider resource development of northern Alaska in particular. As with those Cold War antecedents we detected in chapter 3, aerial technology is providing new opportunities to expand, supply and support resource frontiers.

— 6 —

OPENING UP THE POLES

Our interest in 'opening up' pivots on claims by media and political commentators regarding the rising, possibly even unsettling, interest of South and East Asian states in the Antarctic and Arctic. The New Zealand scholar, Anne-Marie Brady, provides one example when she noted that,

> [m]any observers speculate China's increased polar activities may challenge the interests of other polar states. These concerns are linked to a wider debate about China's international behavior around questions as to whether China is a 'reluctant stakeholder' in the international system and whether China will continue to support international norms as it becomes more dominant. China's polar engagement is a helpful case study towards better understanding Beijing's global behavior and foreign policy' (Brady 2012).

In November 2014, *The Sydney Morning Herald* reported on China's plan to install a satellite facility in Antarctica for its BeiDou satellite navigation system. Although the base – China's fifth – will be constructed under the terms of the Antarctic Treaty, the newspaper noted that it 'heightened concerns about the militarization' of Antarctica and the plan was discussed in terms of how China is 'escalating' its Antarctic involvement (China Daily 2012; Darby 2014).

While China has been the subject of considerable debate and speculation for a decade or so, we explore how such anxieties about 'outsiders' have a longer history. China's northward and southward gaze has provoked considerable concern, even alarm in some quarters, and the country is often reported by journalists and academics alike as being 'resource hungry' for northern and even southern polar hydrocarbons and minerals, as well as ambivalent about international regimes and

142

organizations (e.g. Struzik 2013).[1] China's claim to be a 'near-Arctic state' is treated with scepticism, and concern expressed with the country's global reach and its speculated, but largely unstated, political and economic ambitions (figure 6.1).

The opening words to Stanley Cohen's *Folk Devils and Moral Panics* seem apt for describing such concern: 'Societies appear to be subject, every now and then, to periods of moral panic' (Cohen 1972: 1). Media commentary, for example, on rumours that several thousand Chinese workers are about to descend on Greenland to construct hydropower dams, an aluminium smelter, and the infrastructure for large mining projects, constructs a narrative that often verges close to stylized forms of moral panic. This narrative suggests we (and the 'we' is assumed to be Europeans and North American audiences) should be concerned with recent and emerging global interest in the Arctic, particularly from Asian countries, that implies either a new geopolitical order or a subversion of the assumed political legitimacy of Arctic states alone to govern the region. On Friday, 7 December 2012, Inatsisartut, Greenland's parliament, voted to approve a controversial new law on large-scale industrial projects. Among its provisions, it allows foreign companies to decide on whether they wish to develop projects with imported foreign labour instead of training and employing local Greenlandic people (Nuttall 2012c). In the days and weeks following the law being passed in Nuuk, Greenlandic, Danish and foreign newspapers and other media outlets were awash with reports about an imminent influx of Chinese labourers and claims that the legislation would allow for a process of 'social dumping' in Greenland.[2]

None of these reports make reference to the fact that there are currently no Chinese companies active in Greenland's minerals sector and that there are more Canadian and Australian mining companies involved in exploration and in submitting new applications for development licences. Ironically, perhaps, just before the previous Greenlandic government collapsed in October 2014, there was a planned visit to China so that Greenlandic and Danish officials could

---

[1] Reports abound on how China will overtake the United States as the world's top crude oil importer by the end of the decade. See, for example, Gary Lampher, 'China's thirst for oil growing', *Edmonton Journal*, 22 August 2013, p. D2.

[2] For example, the day the law was passed, Danish broadcaster DR published the following article on its website 'Grønland vedtger lov: nu kommer kineserne' ('Greenland adopts law: now come the Chinese'), available at: http://www.dr.dk/Nyheder/Indland/2012/12/07/195138.htm; while the Greenlandic newspaper ran a similar story in its online edition: 'Nu kommer kineserne' ('Now come the Chinese'), available at: http://sermitsiaq.ag/kommer-kineserne (accessed 26th August 2013).

143

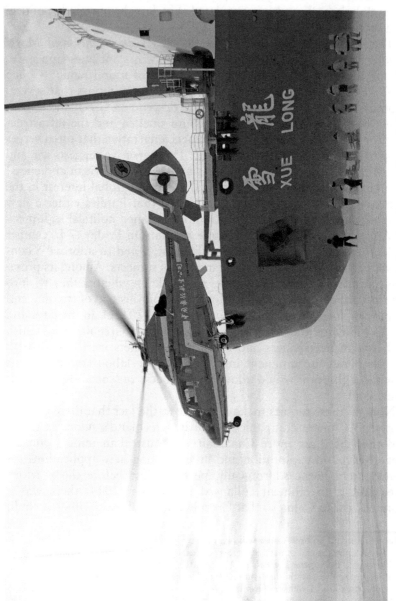

Figure 6.1 The Chinese Icebreaker, *Xue Long* and Ship's Helicopter (Jean de Pomereu)

meet with Chinese and Australian mining investors. It is in this sense that we draw attention to an emergent polar Orientalism, by which we mean a way of representing, imagining, seeing, exaggerating, distorting and fearing 'the East' and its involvement in Arctic affairs. With due acknowledgement to the legacy of the late Palestinian–American scholar and activist Edward Said, polar Orientalism conjures up, obscures and plays with images of Asian political and economic incursions into northern spaces and places, of rich Chinese entrepreneurs attempting to buy large tracts of land in Iceland and Svalbard and/or South Korean universities offering scholarships to Greenlandic students to study the economics of energy, of Inuit communities drawn into business ventures with Asian partners, and of Japanese and South Korean energy companies shipping oil through the melting ice of the Northern Sea Route.

Yet this polar Orientalism obscures and ignores two important aspects of the geography and historiography of the circumpolar North. One is that a considerable area of the Arctic is part of the Eurasian landmass and there are distinctive north Asian qualities to it. We talk about the 'European Arctic', the 'North American Arctic', and 'the Eurasian Arctic', but seldom do we use 'the North Asian Arctic' or 'North Asia' to refer to the Asian portion of Russia lying east of the Urals – this vast region has a population of more than 30 million and is more often referred to as Siberia and the Russian Far East. The second aspect is that Central and East Asian peoples have engaged in historical trade and commerce with Arctic regions and its peoples. Archaeologists have found Chinese beads and pipes in northern Alaska and northern Siberia (Fitzhugh and Chaussonnet 1994), and a seventeenth-century Chinese gold coin in Yukon Territory (CBC News 2011), suggesting indigenous peoples were involved in forms of trade several centuries ago that was far-reaching, while Siberian furs traded in China and Central and Western Asia were an important source of revenue in the eighteenth century (Forsyth 1992).

John Bockstoce (2009) argues that in the nineteenth century the Bering Strait region was the centre of a global maritime trade network connecting the United States, Russia, Great Britain, Hawaii, China, the Philippines and Australia. Iñupiat and Yup'iit along the Bering Strait coast of Alaska were key and proactive participants of this intercontinental trade, acting as suppliers to Russian, American and British traders. They attributed great value to Chinese manufactured trade goods such as glass beads and iron goods, exchanging them for walrus ivory, whale bone and furs from sea otters, beavers, caribou and reindeer, with Chinese people often acting as middlemen in the trade network. Later, Asian gold-seekers crossed the Chilkoot Pass into the Yukon in the late 1890s and an official count in 1901 revealed 84 Japanese, seven Chinese and 'no Hindoo'[3] in the territory. Many stayed and made lives and homes for themselves.

In more recent times, East Asian businesses and commercial activities have been apparent in a number of northern places. Greenland has looked to China and Japan for several decades as important markets for shrimp and fish, while Alaska Native corporations have looked across the North Pacific to South Korea and other Pacific Rim states for economic opportunities. So why the recent worry over East Asian countries 'looking' and 'acting' northwards? Critics might point out that there is a significant difference between eighteenth- or nineteenth-century Chinese desire for sable or beaver furs and what is regarded as today's massive Chinese appetite for energy and resources. China's apparent interest in developing Greenland's iron ore mining potential and extracting its rare earth elements, for instance, has implications for Greenlandic politics and relations with Denmark in terms of sovereignty and security, but it has also made others nervous. But once again Australian or Canadian involvement in Greenland's energy economy barely registers the same kind of commentary as does Chinese interest and potential investment.

So this chapter, as an audit of polar Orientalism, picks up on earlier analyses of orientalism (Said 1978) and what was termed by Morley and Robins (1995) as 'techno-orientalism', a discourse structured by an imaginative geography which positions 'East Asia' and even as we shall see, 'South Asia', as characterized by resource opportunism, selective political networking and strategic obscurantism, coupled with a tendency to lobby for ever greater access to spaces, markets and societies. However, even if unease remains amongst coastal and claimant states in the Arctic and Antarctic, there is also evidence of accommodation and compromise. Arctic states such as Iceland have actively promoted trade cooperation with China and Nordic states more generally have encouraged scientific-intellectual exchanges with their Chinese counterparts. While not undermining what we have termed polar Orientalism, this tells us something about how the Polar Regions are being connected topographically and topologically with East and South Asian states.

---

[3] Yukon Archives (2010) 'Hidden History: the Early Years'. Available at: http://tc.gov.yk.ca/archives/hiddenhistoryasian/en/early.html

The Rise of the East

At the most extreme end of the interpretative spectrum, one finds claims that countries such as China (and to a lesser extent Japan and South Korea) are enrolled in a grand geopolitical strategy designed to rework Halford Mackinder's claims regarding the 'geographical pivot of history'. What appears to be at stake is a new form of geo-political rivalry in the midst of a profound climatic change and geo-physical reshaping of the northern edge of the Euro-Asian landmass and its oceanic region. When China, Japan and Singapore, along with India and Italy, were granted observer status in the Arctic Council in May 2013, many were left wondering whether this signalled the beginnings of a shift in regional Arctic geopolitics. As *The Diplomat* argued:

> China will also now have a secure footing from which it can defend what it will claim to be its 'legitimate rights' in the Arctic. It is quite conceivable that China will now use that foothold to demand as well a voice in the resolution of Arctic territorial boundaries that are up for decision. In 2009–10 it had claimed that no state had sovereignty in the Arctic, a clear slap at Russian claims. Now, to join the Council, it had to repudiate that earlier position and state that it respected the sovereignty of all the states claiming territory in the Arctic but accept that the deci-sion will be made in the future, a sharp contrast to its rigid insistence on its 'core interests' and sovereignty in the Senkakus and the South China Sea. Indeed, given those claims on the seas adjacent to China, it had no choice but to recognize existing exclusive economic zones and bounda-ries if it wanted to be a member of the Council. Nonetheless, it now calls itself a 'near-Arctic state' and an 'Arctic stakeholder'. (Blank 2013)

Writing in 1904, Mackinder claimed that the Central Asian region was pivotal in shaping human geopolitical history because of its central location and access to resources. Since then, generations of writers and political figures have drawn inspiration from Mackinder's speculations about the value of this geographical region, much of which appeared to enjoy a prophetic value when the United States was attempting to 'contain' the Soviet Union (conceived of as an 'Orientalist power') in its 'heartland' and along the so-called 'rim-lands' of Eastern Europe, the Middle East, South Asia and South East Asia.

A decade before Mackinder, however, Charles Henry Pearson published *National Life and Character: A Forecast*. It drew consid-erable attention – it was reviewed at length in *The Sewanee Review* in August 1894 by Theodore Roosevelt, who called it 'one of the

147

most notable books of the century' and urged his political friends in North America and Europe to read it. A former Australian education minister, Pearson warned of the possible effects of increasing Western engagement with Asian societies. China's economic ambitions worried Pearson, and he talked of a future in which Western societies would be 'elbowed and hustled' and even possibly 'thrust aside' by those they had previously looked down upon as inferior and servile, and that the globe would be 'girdled' by the 'Black and Yellow' belt. Asian (and African) societies were on the ascendancy he argued, driven by population growth and industrial capacity. In particular, he argued that this would transform the balance of global power and undermine Western societies, particularly Britain, and he was vehemently opposed to Chinese migration to Australia. Christopher Frayling's account of the rise of Chinaphobia and of the development of the idea that the Chinese were a threat to Western civilization, as expressed in particular by popular culture, throws some light on Mackinder and Pearson's views at the time. In America, Britain and the rest of Europe from the mid-to-late nineteenth century onwards, images of the Chinese as alien 'other' as part of a repertoire of 'Chinese characteristics' grew out of anxieties about such things as immigration, the decline of Empire and the rising dragon. They were, he says, 'about "us" – they were not really about China at all' (Frayling 2014: 10).

In the most recent incarnation, this Mackinder and Pearson-like interest in the extraordinary power (and fear) of particular geographical places has migrated northwards towards the Arctic. And while the theories of race and history underpinning the works of Mackinder and Pearson are absent from ways of representing Asian interests in the Arctic, it is notable how 'population growth', 'industrial capacity' and 'hunger for resources' are common and recurring motifs. Implicitly fusing the perspectives of Mackinder and another early twentieth-century thinker, Vilhjalmur Stefansson (who, for example, predicted in the 1920s that the Arctic would become a 'polar Mediterranean' and was destined to be colonized and settled in the same way as the western prairies of the United States had been in the nineteenth century), Mika Mered of the UK-based consulting firm Polarisk expressed the opinion: 'The Arctic in the 21st century will be the centre of the world. If you control the Arctic, you control the world ... That is the real issue around the Chinese application to the Arctic Council' (Deutsche Welle 2013). The evidence base for such a claim was rather thin in the actual article in question but it does provoke the following question: why would one frame China's

observer application to the Arctic Council in such a manner? The
Chinese were not alone in making such an application, but the can-
didature of other countries did not invite such febrile commentary.
The EU, for instance, had also applied for observer status, but
was rejected (partly, and as we have already discussed, because of
Canadian opposition to EU sealskin directives and other action
affecting Canadian seal hunters and fur trappers). Yet EU ambitions
in the Arctic are seldom reported in terms of dominance and control
(and if anything may well get deferred again due to EU sanctions
against Russia post-Ukraine crisis). In August 2013, the European
Commission's Enterprise and Industry Directorate General put out
a call for tender for a major study on EU needs with regard to cooperation with Greenland. At the heart of the commissioned study
were questions about Greenland's potential for meeting EU needs for
critical raw materials, including rare earths. As the call put it:

> Due to the political and geostrategic importance of Greenland to the EU
> sustainable raw materials supply, the EU has a strong interest in devel-
> oping co-operation with Greenland in this area. Greenland is EU's close
> resource-rich neighbour, especially with regard to rare earths elements.
> Its potential in hydrocarbon (oil) and mineral extraction and processing
> sectors (aluminium, gold, rare earth elements, rubies, uranium) makes
> Greenland an important partner.

Reading the call, it appeared to convey a sense of urgency on the part
of the EU to steal a march on potential rivals by supporting initiatives
to improve and thus acquire geological knowledge of Greenland's
subsurface and to support Greenlandic infrastructure and investment
needs. It is intriguing that China was singled out as eyeing Greenland
more than any other country appears to be:

> The Chinese government is strongly interested in Greenland. Chinese
> President Hu Jintao has just paid a three-day visit to Denmark with
> a large delegation. This was the first visit of a Chinese President to
> Denmark in 62 years. A large London Mining iron ore project which
> aims at selling exclusively to China, a gigantic investment of 2 billion
> EUR is to be constructed in the coming years.

The successful winner of the contract was expected to provide the EU
with a significant report on the availability of Greenland's resources
and reserves, including actual production volumes and forecast of
production volumes in the coming years, analysis of current mining
projects with information about their production potential, incurred
risks and production forecasts per year, the identity of investors,

availability of funding and project risks, and scenario analysis of the development of Greenland's raw materials sector, including estimated production volumes per material and evaluation of risks. Imagine, for the moment, the global media reaction if this call to tender had been put out by the People's Republic of China rather than the European Commission.

David Shambaugh argues that 'China's footprint in the global mining industry is more a matter of reputation than reality' (Shambaugh 2013: 173). He points out that, while the image is one of Chinese mining companies wandering the world and extracting resources, 'the reality is that the vast majority of its imported materials and metals (90 per cent) comes from direct purchases from suppliers or from international commodity markets'. China's control over mines around the world is not currently anything like that of other national governments or multinational corporations. However, it is China's actions over rare earth elements that seem to provoke global reaction over its ambitions for control of valuable resources. China is the world's major producer of rare earths, with some 95 per cent of global production. Yet China has also restricted exports of the rare earth elements it mines. The EU, along with several nations, has complained to the World Trade Organization that such behaviour is discriminatory. No wonder the EU is moving ahead so stridently to consolidate cooperation with Greenland in mapping and mining the subsurface.

Remaking Asian-Arctic Geographies

Political geographer Sanjay Chaturvedi has argued that Arctic geopolitics could best be thought of as a 'historically contingent, but on-going, political project of scripting, staging and projection of the circumpolar northern polar region' (Chaturvedi 2005: 724). Taking his inspiration from critical geopolitical scholarship, Chaturvedi's point, just as we have also explored at length in various parts of this book, is that there has been a long tradition of imagining and representing the Arctic as a series of distinct spaces. Other scholars might use the term polar imaginaries but the point is similar – a repertoire of place-based characterizations and representations (e.g. the Arctic as a frontier space or wilderness, or, from an indigenous perspective counter to hegemonic narratives, a homeland) have proven influential and both durable and resourceful in formulating and mobilizing current and future ambitions and references. The Arctic, in this

150

context, is not a settled unchanging place, but is better understood as an idea and geographical manifestation that has proven dynamic, uneven and contested. As we pointed out in chapter 1, when and where the Arctic begins and ends is a moot point as generations of geographers and others have noted in the past; nowadays the focus is more on the Arctic Ocean (and its connections to global systems of oceanic and climatic circulations) rather than say temperature isotherms, lines of latitude and vegetation patterns. As we explored in chapter 5, while indigenous peoples have fought hard for land claims, resource rights and forms of self-government, the wider recognition of the Arctic as a lived human world that has come out of such grounded struggles seems to be giving way once again to reassembled imaginary geographies of the Arctic as a resource-rich hinterland, a frontier and extractive periphery awaiting exploitation and even plunder from extraterritorial parties (Soikan 2009).

When Mackinder was presenting his lecture, which was later to be published in *The Geographical Journal*, as 'The Geographical Pivot of History', the Polar Regions get a brief mention only. Specifically, the Arctic was described as the 'frozen ocean of the north' and the subarctic as a region where the 'climate is too rigorous, except at the eastern and western extremities, for the development of agricultural settlements' (Mackinder 1904: 423). In his famous map, 'The natural seats of power', the Arctic Ocean is simply labelled 'icy sea', the inference being that mobility at the northern fringes of the 'heartland' is likely to be impossible. Unlike the other oceanic bodies of the world, the shading to the north of the Euro-Asian landmass in particular emphasizes friction and impassability. Clearly, Mackinder did not envisage a commercially or strategically significant northern sea route. But he did sound a warning about East Asian states right at the end of his lecture/paper. As he noted, 'Were the Chinese, for instance, organized by the Japanese, to overthrow the Russian Empire and conquer its territory, they might constitute the yellow peril to the world's freedom just because they would add an oceanic frontage to the resources of the great continent, an advantage as yet denied to the Russian tenant of the pivot region' (Mackinder 1904: 437). Indeed, a few years later, Mackinder wrote that China occupied the globe's most advantageous position, reaching into Central Asia and access to the major shipping lanes of the Pacific (Mackinder 1919).

This idea of the 'oceanic frontage' has proven an appealing one, and might help to explain why countries like Canada and Russia were drawn to the idea of extending their territories, sector-like, all the way to the North Pole, although in Mackinder's essay Russia's oceanic

OPENING UP THE POLES

frontage is blocked by ice and therefore a material barrier to any future geopolitical ambitions. But one might see it as an idea that was fundamentally anticipatory; anticipating a future where access and interest in the North Pole region was likely to change in the midst of exploration, discovery and resource exploitation in higher latitudes. There was always the hope, as nineteenth-century writers expressed, that an 'open polar sea' awaited those who could get beyond the pack ice. Later, other writers were dismissive of ice and weather as any kind of barrier to human innovation, albeit one that was likely to originate from southern constituencies. As Stefansson wrote in *The Northward Course of Empire*, presenting an opposing argument to Mackinder, 'There is no northern boundary beyond which productive enterprise cannot go till North meets North on the opposite shores of the Arctic Ocean as East has met West on the Pacific' (Stefansson 1922: 19). But Mackinder's speculation carried a warning that has been taken up by more contemporary neo-Mackinder observers. James Holmes writing on the *Foreign Policy* blog argued that

> If climate scientists' prophesies of an ice-free Arctic Ocean pan out, the world will witness the most sweeping transformation of geopolitics since the Panama Canal opened. Seafaring nations and industries will react assertively – as they did when merchantmen and ships of war sailing from Atlantic seaports no longer had to circumnavigate South America to reach the Pacific Ocean. There are commercial, constabulary, and military components to this enterprise. The United States must position itself at the forefront of polar sea power along all three axes ... Nations holding waterfront property in the Arctic will bolster their coast guards to police their territorial seas and exclusive economic zones during ice-free intervals. (Holmes 2012)

Warming to his theme, Holmes warned that, 'even partial and episodic access to Arctic sea lanes will add a northern vector to seagoing nations' collective calculus. Not just Arctic countries but countries like China, Japan and South Korea ... will cast their gaze towards such polar entryways as the Bering Strait, Baffin Bay and the Greenland–Iceland–UK gap'.

Earthly shifts are identified in this kind of musing as central to future geopolitical shifts, with a particular focus on those with 'waterfront property' to reimagine their Arctic territories as exposed and even vulnerable to multiple risks and threats. The spectre of an ice-free Arctic Ocean in particular, however uncertain such an outcome might be, has been seized upon by classical geopolitical writers such as Holmes to speculate about an Arctic future where

152

it becomes possible to anticipate new security threats ranging from terrorism and smuggling to extraterritorial resource theft and marine pollution. What unites these items is a fear of flow and movement. At its worst, therefore, it becomes perfectly possible to predict a future for the Arctic, which is fraught with tension and possible conflict. For those closest to the Arctic, indeed those who define themselves as 'Arctic states', there will be ever greater pressure to enhance their constabulary, surveillance and security architectures, the latter term being deliberately used to highlight the role imagination and decision making play in assembling Arctic security complexes (Exner-Pirot 2013).

Two things may deserve further attention here. The first is a geophysical one (the Arctic as 'icy ocean' in Mackinder's terms or even 'polar Mediterranean' in Stefansson's phrasing, albeit writing some 20 years later) and the second a geopolitical one (the geographical representation of global politics). What links the two is a form of naturalized geopolitics in John Agnew's (1998) terms infused with a polar Orientalism that helps not only draw attention to how places get framed and imagined in the first place but the kind of associations and assumptions that get smuggled into such formulations. One such assumption is that East Asian states are increasingly 'mobility rich' when it comes to the Arctic region, using resources, objects and peoples to seize opportunities. Although more contemporary writers do not use the term 'yellow peril', we might ponder whether East Asian interest in the Arctic has reactivated a suspicion of 'their' motives, mobilities and strategies, just as Pearson was writing about in the 1890s. Taking our inspiration from Edward Said (1978) and later Derek Gregory (1994), the focus here is as much on imaginative geographies as it is more material geographies such as the distribution and thickness of sea ice in the Arctic Ocean. Let us start by addressing what appears to be at stake before moving on to what we have termed polar Orientalism and its implications for Arctic and later Antarctic geopolitics, which is as much as discursive as it is embodied and practised. By the end of the chapter, we want to consider how East Asian states and their representatives might actually resist or 'write back' against any lingering forms of polar Orientalism.

Arctic Matter and Mobilities

In his account *On the Move*, Tim Cresswell (2006: 1–2) makes the important point that the idea of mobility is invested with an array of

153

meaning and significance, including links to freedom, opportunity, modernity as well as deviance, resistance and restlessness. Mobility has over the decades been an object of knowledge; something to be analysed and reflected upon by a suite of disciplines including geography, international relations, strategic studies, economics and planning. Ideas about mobility have also been represented in a multiplicity of ways from maps, films, legal documents and images. Mobility is also fundamentally embodied and experiential; our reactions to it vary in intensity and scope. Other scholars such as James Scott (1998) note that mobility was something viewed with some trepidation by state-based authorities eager to impose order and legibility on the communities and landscapes. The transformation of movement to mobility, therefore, was one in which the movement of people, things and ideas becomes transformed into a phenomenon invested with meaning and power. There is always a geopolitics to mobility, therefore. Some people and objects might be mobility-rich while others mobility-poor.

When commentators discuss the prospect of an ice-free Arctic, they are in part referring to a geophysical state change – one which is caught up in a geopolitics of mobility involving sea ice, people, objects and ideas. As climate change scientists and oceanographers have warned, there has been evidence of sea ice retreat (a form of geophysical mobility) since the early 1950s. This has intensified so much that it is no longer unusual to cite minimal sea ice coverage (shrinkage – another form of mobility) in the summer season. In September 2007, a record minimum ice extent was recorded and, as we have discussed, coupled with the Russian flag planting on the bottom of the central Arctic Ocean, this unleashed a series of headlines warning of a new phase of geopolitical intrigue. The geophysical was, it appeared, provoking the geopolitical rather than the other way round.

In 2012, polar scientists recorded that there was evidence of an even bigger reduction of sea ice extent and ever more sea ice was disappearing from the Russian and Alaskan coastlines. Worse still, on both geophysical and geopolitical grounds, multi-year ice was diminishing in significance, and thus thinner year-old sea ice was more vulnerable to melting. The Arctic Ocean, understood with the help of submarine and satellite data sources, was experiencing substantial sea ice reduction both in terms of extent and thickness. In the worst-case scenario, it would appear likely that there might be a period (from e.g. 2040/2050 onwards) between July and November when the Arctic was largely ice-free. Regionally, the thinner areas of sea ice are noted to be around the Russian coastline and the Northern Sea

Route (NSR) but even the thicker ice conditions around Greenland and Canada are being affected by warming trends.

The Arctic Council's Arctic Marine Shipping Assessment (2009) considered the implications that might follow from an increasingly 'ice-free Arctic', and concluded that disputes could follow between states and other parties over the control and regulation of shipping lanes, and maritime and military traffic. The exercise of national jurisdiction by Arctic states such as Canada and Russia was likely to be a source of tension as national regulatory structures and practices coexisted uneasily with the mobilities of extraterritorial parties. Maps and images have played an important role in constructing and circulating the representation of the 'ice-free Arctic', where the frictional costs of mobility are lowered because ice is a diminishing physical impediment. Recent studies have also contributed to this sense of both an opening up and an opportunity for a new kind of geopolitics. In 2013, for example, one study estimated that from 2040 onwards, even open-water vessels might well be able to use the NSR without any need for an ice-breaker escort. With the promise of shorter distances between East Asia and Europe, maritime traffic has grown markedly in the last five years. In 2012, it was recorded that some 46 ships with over 1 million tons of cargo had travelled through the NSR. While the numbers are modest, the maritime traffic trend is likely to continue to be an upward one. While those frictional costs may be the subject of debate, reflecting both uncertainties over sea ice extent, seaborne capacities and commercial costs such as fuel consumption levels, the spectre of the NSR as an opportunistic pathway for both the coastal state and extraterritorial parties is intriguing.

This Arctic mobilities debate, both human and geophysical, is in part one rooted in a concern about the relationship between territory and flows in a neo-liberal world order where state sovereignty remains a dominant idea-practice. But this world order is also haunted by past geopolitical rivalries and speculation over the Euro-Asian landmass. The citation of Mackinder and Spykman reflects that haunting, as if to suggest that the earth's physical geographies continue to have an enduring power, albeit as made clear in the Arctic that geo-power may not be quite so enduring as sea ice melts, thins and shrinks. On the one hand, there are the Arctic coastal states such as Russia and Canada who wish to territorialize the NSR and Northwest Passage and in effect to conceive of them as domesticated spaces under their remit. James Holmes comments on this haunting in explicit terms, 'Receding Arctic ice promises to complete

155

the watery belt enclosing Eurasia – bringing Russian power to the northern frontier of the Heartland at least intermittently. Indeed, the Northern Sea Route passing along Russia's northern coast was ice-free in 2008 for the first time in recorded history (as was the Northwest Passage to Canada's North). Further warming would liberate Russia from its perennial quest for year-round access to the sea while granting the Russian military full use of Eurasia's interior lines. Mackinder's geo-spatial analysis would be complete.' On the other hand, extraterritorial parties such as China, South Korea and Japan (as well as the United States) conceive of an ice-free Arctic as essentially offering opportunities to treat this space as distinctly unexceptional, where mobility is not impeded by over-zealous coastal states. In classical geopolitical terms, this encourages a representation of an ice-free Arctic as an emerging 'rimland' (in Nicholas Spkyman's rather than Mackinder's terms) open for potential exploitation by mobile sea powers. Again in Holmes's (2012) terms, 'Seagoing states should plan ahead to exploit the economic benefits of an ice-free Arctic. Coast guards should ready themselves for police duty in northern waters. But it also befits navies to ponder strategy for this brave new world in the making.' And as Robert Kaplan argued, two years earlier in *Foreign Affairs*,

> The English geographer Sir Halford Mackinder ended his famous 1904 article, 'The Geographical Pivot of History,' with a disturbing reference to China. After explaining why Eurasia was the geostrategic fulcrum of world power, he posited that the Chinese, should they expand their power well beyond their borders ... On land and at sea, abetted by China's favorable location on the map, Beijing's influence is emanating and expanding from Central Asia to the South China Sea, from the Russian Far East to the Indian Ocean. (Kaplan 2010)

The status of shipping routes such as the NSR gets to the heart of this calculus of Arctic geographies and mobilities, as the dynamic geographies and mobilities of the Arctic Ocean provide opportunities and dangers to coastal and non-coastal states alike, depending in part on the distribution and extent of ice and water. Russia, as did its predecessor the Soviet Union, regards the NSR as a vital strategic zone under its sovereign authority. It has historically been hostile to the claim that the NSR was (and ever will be) an international waterway. The United States, by way of contrast, regards the Arctic Ocean and other semi-permanent frozen seas (e.g. the Southern Ocean) as no different than other watery bodies. With the exception of territorial seas, the United States regards the sea as beyond the reach of state appropriation. Limited territorial control by coastal states was strongly tempered by the rights of innocent passage by others. As we also emphasized in chapter 2, the 2008 Ilulissat Declaration reaffirmed the unexceptional qualities of the Arctic Ocean; the Arctic 5 agreed that it was similar to other oceans and as such governed by the provisions of UNCLOS (Dodds 2013).

And yet the Arctic Ocean and other ice-filled seas are acknowledged to possess exceptional qualities and associated challenges. Article 234 of UNCLOS provided extra reassurance for Arctic coastal states such as Russia, with its provision for coastal states to 'adopt and enforce non-discriminatory laws and regulations for the prevention, reduction and control of marine pollution from vessels in ice-covered areas within the limits of the exclusive economic zone … Such laws and regulations shall have due regard to navigation and the protection and preservation of the marine environment based on the best available scientific evidence.' Ice, as Article 234 details, is something that may demand more from both the coastal state and those that attempt to move through it or around it. It is not considered, however, to be an integral element to subsistence lifestyles of indigenous populations in the Arctic. The idea of ice as a hazard prevails, and has implications for how the 'situated knowledge' of indigenous peoples gets marginalized by talk of ship design, navigational costs and the practical authority of coastal states.

This matters because coastal states such as Canada and Russia have used Article 234 to intensify their sovereign authority over ice-filled areas via legislative, constabulary and surveillance-based initiatives and interventions. As Flake (2013: 45) contended, there were severe misgivings about Article 234 at the time of the UNCLOS negotiations precisely because there was a concern that 'coastal states would use environmental protection claims as a pretext to impinge upon freedom of navigation'. For example, Canada passed the Arctic Waters Pollution Prevention Act (extended in 1985 to cover 200 nautical miles from the Canadian baseline) designed to impose greater restrictions on those seeking to navigate through the Northwest Passage. As noted earlier, it has also insisted that foreign vessels participate in NORDREG, a database of all vessels entering Arctic waters in the proximity of Canada. The Canadian government, as part of its volumetric expansionism is invested in a Northern Watch project, designed to help the country literally 'listen' for users of Canadian 'internal waters' via a network of underwater sonar devices. In the context of the NSR, Russia insists upon 'controlling all maritime traffic within 200 nautical miles of its Arctic coastline' (Flake 2013:

157

45). Russia insists that all vessels transiting the NSR have to obtain prior permission from Moscow and follow a series of regulations including escorting. Between 2012 and 2013, the Putin government issued a series of directives, which reinforced these requirements and established a new NSR administrative body to oversee this regulatory structure. Both countries are committed to either making mobility legible in the Arctic seas and/or controlling the size of flow and pace of mobility, especially of commercial shipping.

Growing interest from extraterritorial parties, including those from East Asia, has renewed geopolitical speculation over an ice-free Arctic but also encouraged coastal states such as Canada and Russia to renew their sovereignty-security agendas in the light of concerns that the Arctic is becoming less exceptional. In the Russian context, there has been a public commitment to expand constabulary capacities and the Northern Fleet have commenced patrolling missions, as part of a more intensive monitoring of the NSR. While much attention has been given to subterranean mapping of the seabed and a potential extension of sovereign rights over the outer continental shelf, navigation rights might prove more divisive. If the Arctic becomes less ice-covered, then the right of the coastal state, under the terms of Article 234, becomes less obvious. What may well happen is that coastal states such as Russia and Canada return to a more explicit sectorial geographical imagination to address anxieties about how these countries in the future might impede, slow down or deter foreign vessels, including warships, from entering and transiting its exclusive economic zone waters off its northern coastline. Paradoxically, perhaps, Russia is also keen to remove barriers to improving commercial traffic along the NSR because of the revenue generated by icebreaker escorting and administrative fees.

Diminishing sea ice can reinforce a sense of connection, however. Some accompanying connections may be more welcomed than others. Commercial shipping emanating from ports in South Korea and Japan, for example, provide opportunities for Russian actors to generate additional fee income and exercise coastal state responsibilities. East Asian workers in the service sector economy receive less publicity in places like Svalbard and Greenland, while other East Asian workers in the resource extraction sector provoke fears that towns and villages will be overwhelmed. Even the mobility of pandas, in this case a state-sanctioned transfer from China to Copenhagen Zoo in April 2014, led some to conclude that the deal was related to China's eagerness to invest further in Greenland's mining economy.

Polar Orientalism and the Arctic

The imaginative geographies surrounding the prospect of an ice-free Arctic are central to contemporary manifestations of Arctic geopolitics. As we noted above, what underwrites polar Orientalism is mobility and access. In their analysis of the 'geopolitics of ice melt', Ebinger and Zambetakis (2009) contended that melting and thinning of Arctic sea ice was catalytic, 'transform[ing] the region from one of primarily scientific interest into a maelstrom of competing commercial, national security and environmental concerns, with profound implications for the international legal and political system' (Ebinger and Zambetakis 2009: 1215). This was brought to the fore in the public discussions surrounding East Asian states' interest in the Arctic and their applications for Arctic Council observer membership in the last five years. What is of interest here is not their application histories but rather the reaction that it provoked by simply existing in the first place. In other words, there was a sense of disbelief in some quarters that Asian states would be interested in the Arctic. One such example was provided by *The Economist* which warned in September 2012 that 'Of all Asian countries eyeing the Arctic, it is inevitably China that provokes the most interest and, in some quarters, alarm, for many reasons. It is huge, desperate to secure supplies of energy and other minerals and nervous about the strategic vulnerability implied by its "Malacca dilemma" – that four-fifths of its energy imports pass through that narrow strait near Singapore.'[4]

The 'opening up' of the Arctic is thus a double-edged affair. Geographical features and attributes (including resource potential) are no longer frozen in a way that once impeded movement and access. The melting process is in large part held responsible for the generation of fresh interest from those considered in the past to be remote from the Arctic region. This paradoxical quality could explain why the Arctic Council was not eager to approve observer status for countries like China in 2009 and 2011. In the midst of that uncertainty, there was no shortage of public commentary repeating and reiterating East Asian geopolitical, scientific and commercial interests. According to Linda Jakobson, the 'notion that China has rights in the Arctic can be expected to be repeated in articles by Chinese academics and in comments by Chinese officials until it gradually begins to be perceived as an accepted state of affairs ... There is some irony in the statements of Chinese officials calling on the Arctic states

---

[4] 'Snow Dragons'. *The Economist*, 1 September 2012. Available at: http://www.economist.com/node/21561891

OPENING UP THE POLES

to consider the interests of mankind so that all states can share the Arctic' (Jakobson 2010: 13). The 'rights' refer to non-coastal states and their rights under UNCLOS to the high seas and 'the area', and their rights of innocent and transit passage in and around the Arctic. If Chinese officials are guilty of 'irony' then they would probably not be alone in jealously guarding sovereign rights on the one hand and calling on others to be respectful of the rights of others around the world.

Peter Nolan (2013) argued in a recent essay in *New Left Review* that China's interest in natural resources and the rights of coastal and non-coastal states is often observed through the prism of territorial disputes over the South China Sea, the inference being that either China is attempting to bully smaller regional neighbours and/or determined to extend its extraterritorial influence around the world through strategic partnerships and forthright protection of its rights to the high seas, especially when it comes to fishing. What Nolan notes, however, is that this interest in Chinese behaviour both present (and future) rarely considers how other states, including Britain, France and the United States, often those with a colonial era portfolio of islands, have benefited hugely from UNCLOS entitlements regarding sovereign rights. While China and other East Asian states, as part of their observer status to the Arctic Council, have had to acknowledge the sovereignty (as part of what are termed 'the Nuuk criteria' for observers) of the Arctic states, a lingering suspicion appears to exist about China's motives.

This suspicion took many forms from speculating about why the Chinese Embassy in Iceland is so large to pontificating about what a Chinese Arctic strategy (if it was ever published) might contain. Underlying those interventions lay a concern that it was difficult to detect China's motivations for an Arctic engagement. As Humpert and Raspotnik noted, 'But are geo-economic considerations, especially the access to natural resources and Arctic shipping lanes, the true driver of China's regional policy? China's rapidly growing energy demand and increasing dependence on imports have prompted the country's oil companies to invest heavily in strategic partnerships for overseas oil extraction and production' (2012: 4). The authors go on to conclude that China's partnership with Iceland and interest in Greenland is driven by a regional strategy designed to consolidate China's status as an emerging global power. Intriguingly, their report is labelled 'From Great Wall to Great White North', which, apart from being judged to be 'eye-catching', is not explained to the reader. We are instead left to infer that Chinese strategy is inherently expansionist and opportunistic

160

rather than defensive and modest. Other observers, such as Anne-Marie Brady have been scathing of Chinese investment in Arctic science: 'Beijing produces a lot of smoke, mirrors and big talk, which disguises their small investment' (Brady 2012). This in turn invites further speculation about what China is really interested in when it comes to the Arctic. She concludes that China's polar behaviour is a good guide to the real drivers of Chinese behaviour – status anxiety, resource acquisition, strategic partnerships and geo-strategic projection.

Edward Said (1978) made the important point about the way in which colonial powers mapped their world onto the worlds that they encountered, pacified and administered. What was at stake was not just a material conquest (e.g. troops on the ground, the imposition of infrastructure and the like) but also an imaginative one; a way of reading those colonized landscapes, including its peoples and categorizing and ordering in the process. Said's focal point was the Napoleonic invasion of Egypt in the late eighteenth century and the onset of a new round of colonial-imperial expansion, which sought to make Egypt and the wider Middle East accessible and legible to European and later American audiences. Orientalism was as much a system of representation as it was a set of performances taking the form of books, museums, paintings, and later film and television. The discourse of Orientalism was inherently performance-led; it helped to create the very thing 'the Orient' that it named. In this context, 'the Orient' was understood as an exotic, even bizarre, space that unsettled European and later American sensibilities. Coveted on the one hand and yet considered monstrous on the other. Later, Timothy Mitchell drew attention to how colonial Egypt was to be made legible and subject to the calculations of colonizing power (Mitchell 2002).

The idea of 'the Orient' being orderable and calculable made more likely that colonizing powers could indeed affect change and address the monstrous and disorderly qualities of places like Egypt. Said, and later Gregory, note that this ordering impulse was strongly linked to militarism and politico-economic regulation. While the political-strategic was one register, another was the commercial as European colonizers saw the Middle East as a space for trading and exchange. But such possibilities also carried dangers. A contemporary of Mackinder, the American naval office and writer Thomas Mahan warned that the Middle East was imperilled by a toxic cocktail of German, Ottoman and Russian geopolitical ambition. As other scholars have noted, British governments used these apparent machinations to justify their own interventions, including drawing new boundary lines on the post-World War I map.

161

What relevance might this have for the contemporary Arctic? We have used the term polar Orientalism as a way of reflecting on the multiplicity of ways in which others have imposed their own ideas and interests on the Arctic, but also the way in which those who come from outside the Euro-American world are positioned within Arctic geopolitical discourses. So while there is more to be said on indigenous politics and representations, our focus remains on a different category of observer/participant; East Asian states. One important element to the 'disciplining' of these East Asian states, especially China, was through their application to be observer states to the Arctic Council. As Stephen Blank noted,

> Now, to join the Council, it had to repudiate that position [of not accepting the Arctic sovereignty of the eight states] and state that it respected the sovereignty of all the states claiming territory in the Arctic but accept that the decision will be made in the future – a sharp contrast to its rigid insistence on its 'core interests' and sovereignty in the Senkakus and the South China Sea. Indeed, given those claims, Beijing has no choice but to do so. Nonetheless, it now calls itself a 'near Arctic state' and an 'Arctic stakeholder'. (Blank 2013)

So perhaps the focus should not be on how the 'Arctic' is represented per se but how new 'observers', especially those hailing from East Asia, are being encouraged to make sense of this region and its inhabitants in a way that is respectful of the fixed sovereignty of the eight Arctic states. East Asian states are reimagining themselves as geographically proximate to the Arctic region, as geophysical change and increased movement of people are creating opportunities for investment and even settlement. These activities and imaginaries generate uncertainties for Arctic states as they expect to be included in political decision-making.

Polar Orientalism and the Antarctic

While much of our focus thus far has been on the Arctic, it might be useful to interject a southerly counter-point. How have seven claimant states such as Australia, New Zealand and the United Kingdom, rather than internationally recognized Arctic Ocean coastal states and other Arctic states responded to 'Asian' interest in the Antarctic? Rather than start with China, we might turn our attention towards India. In 1956, India made a formal proposal that the United Nations consider the 'peaceful utilization of Antarctica' and posited the claim

162

that, 'the mineral wealth of the landmass is believed to be considerable and its coastal waters contained important food resources'. This expression of interest was not welcomed by India's Commonwealth allies Australia, New Zealand and the UK, who feared that this proposal was the start of a push by newly independent countries to explore a new regime for the Antarctic, irrespective of existing sovereign claimants. After intense lobbying, India was persuaded to drop its proposal for a form of international trusteeship in Antarctica.

Indian interest in the future of Antarctica emerged again in 1958 and again provoked intense speculation as to what might be motivating such interest. For established Antarctic states such as the UK, there was genuine confusion as to what might be at stake; was it concern over nuclear and military activity in Antarctica? Was it the resource potential of Antarctica? Or was it animated by a desire not to be excluded from any future political regime that might govern the Antarctic after the International Geophysical Year? India was not alone in raising the last question, as the Soviet Union made clear to the United States that it was not going to be excluded from any treaty or regime negotiations regarding Antarctica.

While India was not an original signatory to the Antarctic Treaty (1959), its attempt to raise the issue of the future governance of Antarctica did have an impact on the treaty negotiations and eventual treaty itself. All the original 12 signatories recognized that the Antarctic would have to be demilitarized and denuclearized, and the question of resource access was not considered. However, in a blatant attempt to regulate the application of new members there was provision that ensured that only those who engaged in 'substantial scientific activity' could become a full consultative party. India's intervention in 1956 forced the UK government to consider whether offering India a 'second tier' membership of the Antarctic Treaty regime might not be strategically wise. This was withdrawn later as concern was expressed that other African and Asian countries might want to join as well.

Until the early 1980s, India and other Asian states such as China were not actively involved in Antarctic politics and science even if they were concerned with access to the resources of the Southern Ocean. It is worth bearing in mind that India, as a member of the Non-Aligned Movement, was supportive of new proposals in the 1970s to explore how the global community as a whole might manage Antarctica's resources. Prime Minister Indira Ghandi reminded her audiences in the early 1980s that the Indian Ocean linked India to Antarctica, as preparations began in earnest for the first Indian expedition to the far south in 1981.

163

India's first expedition led to the establishment of its first research station and the start of India's direct engagement with Antarctica. As with earlier European and North American explorers, flags and inscriptions were left on the ice and subsequent interest mounted in being involved with the Antarctic Treaty System, as it negotiated living resource management. But Indian scientists also began to investigate the geological and meteorological connections between Antarctica and India, and their research proposals echoed earlier work by southern hemispheric states such as Argentina, Chile, South Africa, New Zealand and Australia, all of which have imagined themselves as 'gateways' to the Antarctic and Southern Ocean, as well as being fundamentally connected to those southerly spaces.

The ATS was fundamentally transformed when India and China were admitted, with some haste, as consultative parties. The decision to embrace both Asian states was unquestionably driven by international political pressures, as the Non-Aligned Movement and UN members led by other states, notably Malaysia, pressed for greater accountability and challenged the right of the ATS to act on behalf of the wider international community. All of this was further heightened in the 1980s as the ATS found itself under greater scrutiny and criticism from environmental groups, who took issue with the mineral resource regulations negotiations (CRAMRA 1982–8). Throughout that decade and beyond, the ATS membership was transformed by the influx of new members from Asia and Latin America.

China, South Korea and India's base building programmes have been cause for some alarm as they increasingly make their presence felt on the polar continent. China has established four research stations and plans a fifth, leaving commentators in Europe, North America and Australasia to reflect on not only the polar research budgets for Asian Antarctic activities but also the long-term implications of this investment. As *The Economist* opined in November 2013,

China is steadily implementing its considerable polar ambitions. Over the past two decades its yearly Antarctic spending has increased from $20m to $55m, some three times the country's investment in the Arctic. There are many reasons to stake a claim, not least to bolster national pride and global geopolitical clout. The goal of the current five-year polar plan, according to Chen Lianzeng, the deputy head of China's Arctic and Antarctic administration, is to increase the country's status and influence, in order to protect its 'polar rights'.[5]

5 http://www.economist.com/blogs/analects/2013/11/china-antarctic

One interesting reaction was to be detected in Australia when it was announced that China had established a station in 2008 at Dome A, an area in the 'Australian Antarctic Territory' that was so remote that it took repeated attempts to reach that part of Antarctica. The underlying reaction appeared to be one of shame and disappointment that Australia was not in the position to claim that particular Antarctic 'first'.

A year before, two Australian commentators Anthony Bergin and Marcus Haward released a pamphlet entitled 'Frozen Assets: Securing Australia's Antarctic Future' (Bergin and Haward 2007), which seemed to coincide with a broader exhibition of what was termed 'frontier vigilantism' (figure 6.2 and Dodds and Hemmings 2009) – an imaginative geography that positioned Chinese, Indian and South Korean Antarctic activity as worrisome and in the Australian case, threatening to their interests in Australian Antarctic Territory. Under the terms of the Antarctic Treaty and associated legal instruments such as the Madrid Protocol, the geographical location of research stations is a matter of scientific and environmental impact assessment. Claimant states cannot, for instance, object to the construction of a new base because it happens to be in their claimed territory.

Figure 6.2 Zhongshan Research Station, Antarctica
(Jean de Pomereu)

165

However, Australia and other claimants have clearly watched in alarm as larger states such as China have located their research stations around the Antarctic in a manner reminiscent of the United States and Russia/Soviet Union – in other words, a form of what we might think of as colonial mimicry. China, India and South Korea all participate in Antarctic place naming and use exactly the same sort of nationalistic, patriotic, commercial sponsorship and symbols that European and North American states did in the nineteenth and twentieth centuries. In the Australian media, there is no shortage of articles complaining that the AAT is vulnerable to this external intervention and that the consensus-based ATS is insufficiently robust to challenge such opportunism on the part of newer members. The 2014 Australian Senate Committee's report on Australia's future role in Antarctica and the Southern Ocean recommended further investment and involvement in the region. The report also reflected on what it termed 'emerging players' and noted:

> Much has been written and said in recent years about the increasing interest of 'new players' in the Antarctic region. A number of emerging nations including China, India, Malaysia and the Republic of Korea are rapidly increasing their investments and activities in the region, giving rise to speculation about the nature of their interests, and concern about the declining influence of the traditional Antarctic powers. In particular, the growing profile of China as an Antarctic actor was mentioned frequently to the committee. China joined the Antarctic Treaty in 1983, but its engagement was relatively modest until this century. In the last ten years China has significantly increased its investment in the Antarctic region, including more than doubling spending on Antarctic science and logistics, and building new bases on the continent itself, including in the Australian Antarctic Territory.[6]

The concern over investment, spending and interests has also manifested itself in debates and controversies regarding marine conservation. Joined by other countries such as Russia and Ukraine, Australia, New Zealand and other original signatories to the Antarctic Treaty routinely note their concern about the reluctance of Russia and China to agree to the establishment of marine protected areas. A longer standing antagonism exists between Australia and New Zealand with Japan and its 'scientific whaling' in the Southern Ocean. Russia and China

---

[6] Australian Senate Foreign Affairs, Defence and Trade References Committee Australia's future activities and responsibilities in the Southern Ocean and Antarctic waters (October 2014). Available at: http://www.aph.gov.au/Parliamentary_Business/Committees/Senate/Foreign_Affairs_Defence_and_Trade/Southern_Ocean_and_Antarctic_waters/~/media/Committees/fadt_ctte/Southern_Ocean_and_Antarctic_waters/report/report.pdf

have been accused of being intent on prospecting for mineral resources, despite the prohibition contained within the Madrid Protocol. What is apparent is that the status accorded to the value of consensus in the Antarctic Treaty System is changing – in previous generations 'consensus' would have been viewed as something to be celebrated; now it is more likely to invite cynicism about how countries can exercise a 'veto' and thus prevent consensus from coagulating.

All of the above feeds a southerly polar Orientalism, which is fundamentally suspicious of East Asian states, Ukraine and Russia and their motivations for being involved in Antarctic and Southern Ocean activities. Aspiring countries such as Belarus and Iran feed further into that fearful imagination which posits the question – what do they want from Antarctica? While at the same time, claimant states such as the UK have no difficulty in renaming a large part of the Antarctic Peninsula 'Queen Elizabeth Land' in December 2012 in an apparent tribute to the Queen's fiftieth anniversary as sovereign of the United Kingdom and Commonwealth. Other claimant states have long taken an interest in resource management, and sought preferential access to fishing grounds in the Southern Ocean. So while Europeans and Australasians are urged to remain vigilant, we might wonder how unique East and South Asian states are when it comes to shoring up their interests in Antarctica. If anything, those newer members have simply copied the example of older member states, the difference being that their science budgets in the case of South Korea now exceed that of the UK, while India carried out its own expedition to the South Pole in 2010.

Managing Polar Orientalism

This final section briefly considers how East Asian states might resist such analyses of their motives, intentions and strategies, and explores how these states and associated organizations may have managed lingering. Unlike for the Antarctic, there are no Article IV (which freezes existing sovereignty positions) arrangements in the Arctic which means that sovereignty has a very different legal and political (even visual) purchase in this region. So, as discussed, China, South Korea and India have been largely free to invest in scientific stations, infrastructure and engage in place naming and other European–North American forms of polar nationalism in the Antarctic. Acknowledging (and being seen to be doing so) the more sensitive and immediate issue of Arctic sovereignty (and the Arctic 8 as a whole) was an important and necessary measure for their acceptance as observers to the Arctic Council in May 2013.

One decision taken by some of the East Asian states has been to appoint Arctic ambassadors. Japan appointed such a figure in March 2013, an intervention deliberately timed to highlight its seriousness as the May 2013 Arctic Council ministerial meeting approached. Another strategy was to highlight participation and attendance at Arctic Council meetings, as if to suggest that this 'commitment' proved that these observer candidates were both serious about their Arctic interests but also respectful of the Arctic Council as an intergovernmental forum. Japan, for example, was careful to list its participation in such ministerial and expert meetings as well as noting the creation of a Consortium for Arctic Environmental Research. A third was to lobby member states, especially the Nordic contingent, in the run-up to the 2013 ministerial meeting in northern Sweden. Finally, China, Japan and South Korea in particular have made great play of their polar capabilities, whether it be in shipping or in scientific terms. South Korean investment in polar science has been noted with particular reference to new investment in the Korean Polar Research Institute and new polar research vessel, *Araon*.

What is striking in the Chinese case is how the idea of China being a 'responsible' state is being mobilized as a way to manage fears and anxieties about its longer term intentions in the Arctic. This involves a repeated acknowledgement of the Arctic sovereignty of the eight states in question, while at the same noting that the Arctic as a region is of interest to a wider community of states including China. However, Chinese media do report on 'incidents' which highlight that China's interests in the Arctic are scrutinized more heavily than other countries such as Japan and South Korea let alone Britain and Germany. One interesting example was the high profile rejection of a Chinese company's proposal to purchase land in Iceland for an eco-tourism project. The rejection of this plan was blamed on racism and discrimination. As the Chinese news agency, Xinhua, reported in November 2011:

> A Chinese businessman criticized Icelandic authorities' rejection of his land purchase plan in the country, saying the decision highlighted the prejudice faced by Chinese investors abroad … [The businessman] blamed the western countries for imposing 'double standards,' saying they are eager to 'encourage the opening of the Chinese market while they close their doors to Chinese investments.' The issue stirred controversy as some media reports hinted that the proposed investment could provide a cover for China's geopolitical interests around the Arctic …

168

The rejection shows a continuity of the Cold War mentality that 'invest-ment from private Chinese entrepreneurs is a threat to national safety'. (Xinhua News Agency 2011)

Linda Jakobson notes, 'China's Arctic aspirations, in particular, have evoked concern, even anxiety, that throughout history has accom-panied the rise of a large power' (2012: 3). Significantly, individual business people are bound up in this narrative and then taken as rep-resentative of the country itself.

In spite of the strategies/efforts mentioned above, East Asian states, China, Japan and South Korea, find themselves in something of a bind when it comes to organizations such as the Arctic Council. How to demonstrate one's Arctic credentials (respectfully) without risking accusations of mobilizing wider geopolitical/geo-strategic agendas? The Arctic Council's admittance of these states and others such as Singapore and Italy is simply an acknowledgement that their pres-ence was too great to avoid, especially if one recognized that these states and the actors associated with them are playing their part in shaping and refashioning Arctic territories, whether it be through South Korean ship building, Japanese scientific investment, Indian research activity in Svalbard or immigrant workers from China. East Asian states have a territorial stake in the Arctic as well as a relation-ship based on flow and mobility. The non-Arctic states are asserting a degree of rootedness by calling themselves 'near-Arctic' (as China does) or by establishing research stations on Svalbard (China, Japan and South Korea), where the special conditions of Norwegian sov-ereignty permit a high level of international presence. But as noted, there may be other less visible ways, at least to Arctic states and their political representatives, that East Asian citizens are making their own presence felt. Non-Arctic states may well be more interested in establishing routes (to wealth, as well as through the Arctic's waters) than in building on existing foundations in the region, but when they do seek a more permanent presence, this can and does unleash ori-entalist discourses and practices; predicting Arctic futures in which extra-territorial others exploit resource wealth, unsettle sovereign presence or exert strategic presence.

Conclusion

The growing interest of East Asian states in the Arctic, in particular, has attracted considerable commentary and reflection. For realist/

169

geopolitical writers this interest and investment signalled a future that was likely to involve competitive behaviour, as countries such as China were complicit in a 'new great game' of resource extraction and military projection. More liberal writers turned to intergovernmental forums such as the Arctic Council and argued that through the admittance of China and other East Asian states as 'observers', it was possible to accommodate competing interests. The future of the Arctic Council itself has often been the focus of analytical attention; a prism for discerning likely Arctic futures. This might prove all the more interesting in the midst of a Canadian chairmanship (2013–15), especially in a context when Canada, in particular, has repeatedly emphasized its local and inhabited identity (through the presence of indigenous people 'since time immemorial') as an Arctic state and circumpolar actor. Institutionally, this might create further pressures on the Arctic Council and its collective imagination of an Arctic as a grounded rather than routed space. Maybe one outcome in May 2013 was to reinforce a membership where states in the form of permanent observers take precedence over non-state members, thus ensuring that the interests of distant investors, resource corporations and shipping companies are channelled through state representatives. This appeared to nip in the bud a rival initiative called Arctic Circle established in 2013; led by the President of Iceland, it promised to be an 'open tent' for Asian and other stakeholders.

Arctic states, and their representatives, are increasingly invested in sovereign space-making practices, which in turn are increasingly invested in focusing on reinforcing modes of governance and institutional patterning. Charting and auditing East Asian interest has played a vital role in this sovereign spatialization of the Arctic; monitoring investment in science, trade flows, resource extraction, foreign direct investment, trade deals and so on. China's engagement with Iceland, the Nordics and Canada over scientific (e.g. the new China-Nordic Arctic Research Centre in Shanghai) and energy matters remind us that China is very much sourced in the Arctic and other northern places such as the oil sands in Alberta. While there has been no shortage of large-scale posturing regarding a Mackinder-like struggle for the Arctic (and even the wider northern fringes of the Euro-Asian landmass), we might also ponder whether there are also more subtle and small-scale interventions which are playing their part in remaking the Arctic; a region where territories and flows are reconfigured as much by sea ice as by the migration of East Asian service workers to hotels in places like Nuuk and Longyearbyen. And longer term change in the central Arctic Ocean might well open up further opportunities for a raft of actors including China and the EU to make their fishing presence felt in the high seas.

But let us end with the other polar region – the Antarctic. In December 2013, the Chinese icebreaker *Xue Long* (the Snow Dragon) affected a rescue mission of a Russian registered vessel chartered by an Australian expedition team intent on recreating the trials and tribulations of the Australian Antarctic Expedition of 1911–13. Australian and French operators tried to rescue the ice-bound Russian vessel but without any success. The *Xue Long* diverted from its own expedition to initiate China's fourth Antarctic station. Using its helicopter, the Chinese crew air-lifted the stuck passengers and transferred them to their own vessel. Later an Australian vessel, the *Aurora Australis* returned the Australian expedition members to Hobart. The rescue operation by the Chinese illustrated the double-edged nature of fears of Chinese, and for that matter South Korean, investment in science and infrastructure; these countries have both contributed to search and rescue operations. Perhaps it is the sign of the times. Chinese sailors and airmen saved an expedition designed to revere the Anglo-Australian 'Spirit of Mawson'.

— 7 —

POLAR DEMANDS AND DEMANDING POLAR REGIONS

Our account, and audit, of the contemporary geopolitics of the Polar Regions is inspired by a counter-reaction to the febrile reporting, speculation and imagining about the futures of the Arctic and Antarctic in the last decade or so. In that sense, the media – as well as a considerable academic – response to the planting of a Russian flag on the bottom of the central Arctic Ocean has much to answer for in terms of stimulating an upsurge of what we have described as scrambling discourse. It would, however, be naïve to hold culpable one image of a submersible depositing such a flag on the seabed for current global interest in, and concern about, such scrambling for territory and resources per se. While we have pointed to a range of drivers that we believe responsible for scrambling the Polar Regions, we also want to remind our readers that we should not over-estimate novelty either. Elsewhere, in both time and space, others have declared that a new scramble for the Arctic and/or Antarctic was imminent. In April 1929, for example, the *Adelaide Advertiser* suggested that there was a 'scramble for the Antarctic' in the wake of concern that commercial whaling combined with international geopolitical interest in the Southern Ocean was intensifying (cited in Griffith 2007: 111). Decades later, K.S.R. Menon warned of a new 'scramble for the Antarctic' in the wake of negotiations designed to formalize an international legal regime for the regulation of mining on the continent (Menon 1982).

In this final chapter, we reflect on how these scrambles and scrambling in general contribute to the making of a 'demanding geopolitics' both past and present. These are regions that have not only demanded attention as we saw with regard to nineteenth-century contemporaries but are making demands of their own in the twenty-first century,

172

as frontline areas of climate change, resource exploitation, indigenous rights, and the interaction between the exclusive rights of sovereign states (and claimant states) and those belonging to the international community. That climate change places demands on infrastructure and communities, and that demands are made to resolve the challenges that arise, is one of the most striking and obvious illustrations of this. In the circumpolar North, a distinct warming trend in lowland permafrost of 2–4°C over the last hundred years, with thawing, thermokarst formation, and severe erosion, not only disturbs animals and vegetation growth seasons, but has its consequences for everyday human life. Warming is likely to increase the possibility that people will face a range of natural hazards as part of daily life (the risks involved in travelling on land and sea, for example, or exposure to ultraviolet radiation), will demand action to respond effectively, or require different skills in navigating shifting terrains and volatile waters. In northern Greenland, unstable sea ice is beginning to make ice-edge hunting more difficult and dangerous, placing demands on knowledge, experience and skill. Hunters report anxious feelings when travelling on ice they say is now often more slippery than they have experienced it, while changes in snow cover are also causing difficulty in accessing hunting and fishing areas by dogsled or snowmobile, making necessary local adjustments in winter travel, hunting practices and fishing strategies (Nuttall 2009). The widespread thawing of discontinuous permafrost in Alaska demonstrates some of these hazards and the implications for both habitat change and the physical infrastructure of rural and urban communities. In western Alaska several communities in low-lying coastal and island areas, including Shishmaref, Kivalina and Little Diomede have been affected significantly by recent climate events and changes and face severe problems due to coastal erosion and thawing of discontinuous permafrost, as well as increased storminess. The Iñupiat village of Shishmaref, located on Alaska's Sarichef Island in the Chukchi Sea, is home to some 560 people who rely on a subsistence lifestyle of hunting and fishing. Coastal erosion and the frequency of storms have forced the residents to vote in favour of relocating the village to the Alaskan mainland (Marino 2012). But such change also places demands on ecosystems – scientists say that permafrost throughout the Arctic will thaw more quickly in spring, but take longer to refreeze in autumn, and permafrost boundaries will gradually move poleward, with most of the ice-rich discontinuous permafrost disappearing by the end of the twenty-first century. The tree line will move further north concurrently. The dark evergreen forests characteristic

173

of the boreal regions south of the tree line are beginning to advance northwards into areas where cedars, firs, pines, spruces and birches have not been known to grow.

The Polar Regions have also demanded particular kinds of approaches to living in them, developing them, extracting resources from them, and to studying and researching them. These approaches have, as we discussed in earlier chapters, produced particular kinds of histories and narratives about the Arctic and Antarctic as extraordinary or exceptional places. Development and the construction of northern cities, mines, pipelines, airbases and so on, have often been celebrated as a triumph over inhospitable natures that make impossible demands on people and equipment. But in the process of incursions and movement into polar places, lands are conquered and settled, indigenous people are settled, 'civilized' and their lives modernized, and the frozen parts of the planet reveal their mysteries and give up their secrets. Exploration, discovery and industrial ventures have often been framed in language that suggests an assault on the unknown, the harsh, the unforgiving and the primitive, demanding the deployment and use of what Paul Josephson calls brute force technology (Josephson 2002); exemplified by the mobilization of labour and machinery to carve Camp Century from Greenland's inland ice, for instance, or the construction by the Soviets of large resource towns in the Russian North. H.P. Smolka begins his book on Russia's polar empire, *Forty Thousand against the Arctic*, as follows:

> There are two maps in this book – one at the beginning, one at the end. They belie each other. One pretends that the whole northern edge of Asia, the Arctic, cannot be developed. No civilized people (it declares) can live there regularly. Its earth cannot or will not yield its treasures to mankind; modern travel is impossible. A death sentence is put upon a continent. The other says it can be done. There can be life. Ships can travel along the shores during part of the year; they can be guided with ice breakers. Aeroplanes can link the centres of man and keep up contacts throughout the year. Plants can be made to grow on the soil. Precious minerals can be lifted from the earth, factories constructed, towns built. Thousands can live there. (Smolka 1938: 15)

While cities were built on the tundra to house people who worked in mines, factories and smelters, other Soviet plans to transform the Arctic were not realized, such as projects to divert rivers that flow into the Arctic Ocean and build a dam across the Bering Strait. Understanding the history of these scrambles for resources and ambitions for large-scale development is important given the Arctic is

174

undergoing a process of re-frontierization – as such, the region is re-imagined as a place of opportunity in which new kinds of technology need to be developed and deployed to extract resources, while new designs for northern living are considered critical for societies under rapid transformation. As the conference series 'Future Greenland', an event held in Nuuk focusing on development and business opportunities, illustrates, there are visions of a new Arctic demanding training, skills and innovation in cold region engineering, architecture and commerce.

Our concluding focus on demands and demanding is a deliberate one, designed to highlight three themes. The first concerns the manner in which demands for action and inaction impact and shape the contemporary geopolitics of the Polar Regions. We might want to consider as part of that those moments when things, histories and geographical connections are ignored and marginalized. Sometimes we do not want to listen to, see or let alone feel 'demands'. At times, however, we have no choice but to consider the demands of others as Lego discovered when Greenpeace repeatedly campaigned against their product association with Shell in 2014. While Lego initially tried to ignore such campaigning, they eventually announced that their partnership with Shell would cease after relentless criticism of Shell's drilling plans in Alaska.

The second is to consider how demands recalibrate our sense of proximity and distance. As we asked at the beginning of this book, where does the Arctic and Antarctic begin and end? For countries such as Argentina and Chile, there has been a long-standing belief that they are intimately connected to the Antarctic Peninsula. Geological and geographical connections matter greatly, and generations of Argentine and Chilean school children have been taught to understand how their countries extend all the way to the South Pole. For both these countries, the presence of the UK as counter-claimant and occupier of the Falklands and South Georgia is an awkward reminder of an artificial proximity of a mainland state located some 10,000 miles away from Antarctica.

The third theme is about how demands challenge us to think about our relationship and responsibility towards others, both human and non-human actors; the people and communities who live in the Polar Regions but also the polar ice sheets and seals, whales, polar bears and reindeer. In the Arctic, indigenous peoples demand to be 'heard' and have become ever more active in articulating their demands for autonomy, resource rights, participation in decision-making processes about the conservation of animals, lands and waters, and recognition from others that they are rights holders rather than stakeholders, thus bringing to the fore a marked distinction between indigenous communities and their relationship to northern places as homelands and those from outside the Arctic who claim vested interests in the region. The rights of Greenlanders and Canadian Inuit to hunt polar bears, for instance rub up against the campaigns of environmental groups to protect them from being hunted in the first place. Our relationship to distant others, as evidenced by the growing interest and involvement by extraterritorial parties in the geopolitics of the Arctic and Antarctic, is another aspect to this that can and does provoke awkward, angry and fearful reactions.

Demanding Polar Geopolitics

The demand for action, in its broadest sense, has been a defining feature of contemporary geopolitics of the Polar Regions. In 2014, for instance, we witnessed celebrities such as British actor Emma Thompson holding up placards demanding that the 'Arctic' be saved from further industrial exploitation. Sponsored by Greenpeace, she was enrolled to support their 'Save the Arctic' campaign, which hopes to accrue 10 million signatures in an online petition demanding that the area around the North Pole be secured as a global sanctuary and that further oil development and commercial fishing be prohibited in the Arctic. The photographic images of Thompson and her placards saying 'Save the Arctic now' and 'Where I stand should be ice' tell us something about the visual and indeed affectual power of the photograph (see figure 7.1). Campaigning groups make widespread use of such visual images, especially of polar bears roaming on increasingly thin sea ice, and act as moral entrepreneurs as they seek to gain our attention, appealing in particular to those who will never visit the Arctic and may be located thousands of miles away.

The photographs of Emma Thompson and her placards act as a form of visual agitation (the photographs hope to provoke action rather than despair) and we would contend geopolitical and biopolitical simplification. Her body and her placards dominate the photographs in question. The Arctic is not easily identifiable and there is never any acknowledgement that the Arctic might be an inhabited space. The only people and objects that feature in the photographs released by Greenpeace are her accompanying party and the Greenpeace ship *Arctic Sunrise*. While these photographs are intended to provoke attention and to make demands on our emotions

Figure 7.1 Emma Thompson Making 'Demands' of the Arctic (Cobbing/
Greenpeace)

(and presumably encourage viewers to sign the online petition), they
also remind us of what is absent and how that might unwittingly
perhaps resurrect colonial ways of seeing the Arctic, where white
men and women prefigure in landscapes and seascapes 'emptied out'
of indigenous peoples and northern communities. This campaign is
reminiscent of the use of actors such as Brigitte Bardot in 1977 and
musicians such as Paul McCartney in 2006 by an animal rights group
and the anti-sealing lobby in their very public opposition to seal
hunting off the coast of Newfoundland and Labrador. For northern
sealing communities, such images act as reminders of the damage
inflicted on their livelihoods and economies and the campaigns that
continue to haunt them.

This kind of activity perpetuates a highly ethnocentric view of the
Arctic in which it takes 'distant others' such as Thompson to 'represent' it to wider global audiences rather than those who actually live
in those places. But this is not new of course. Countless explorers
and adventurers used their sketches, photographs, maps, charts and
the material objects of indigenous communities to inform and enter-
tain European-based learned societies and public audiences. While
Internet communications have altered the speed and volume of the
mobility of images and texts, earlier images, novels, paintings and

maps attempted to generate a more intense sense of proximity for distant Arctic ice in places like the UK, while paradoxically distancing those same communities from the realities of the Arctic region, including long-settled indigenous and northern communities with their own set of demands.

One current example of the efforts of international environmental organizations to protect the Arctic is WWF's Last Ice Area initiative. As the website for the project declares, 'This region in the high Arctic of Canada and Greenland was identified using cutting-edge sea ice models, and it's projected to be the last stronghold of summer sea ice as the Earth continues to warm due to climate change. In the coming years, it will be essential as an enduring home for wildlife, and therefore for the communities that depend on those species.' But this place represented as the last ice area is a human world of Inuit communities in which the movement of animals, the advance and retreat of glaciers, the dripping and trickling sounds of meltwater, the texture and distribution of thinning ice, the increasing frequency and ferocity of storms, and the forces of coastal erosion are considered to mean something powerful and extraordinary, as well as troubling and worrying, to those who live there, indicating the sentience, agency and predictability of a world of becoming (Nuttall 2009). While WWF makes an appropriate nod in the direction of Inuit communities, the Last Ice Area initiative sets out to privilege the protection of the environment and animals such as the polar bear over the interests of those who live and hunt there.

The manner in which Arctic environments, lives and animals are represented and acted upon is deeply controversial given a prevailing context where seal product bans cause great resentment within communities in Canada and Greenland (e.g. Wenzel 1991; Lynge 1992). In June 2014, the Commission for Environmental Cooperation voted 2–1 to reject a request filed by the US Center for Biological Diversity to investigate why Canada has chosen not to designate polar bears as threatened or endangered as a result of sea ice loss and projections of declining numbers. The commission was established as part of the North America Free Trade Agreement as a way of monitoring pressure placed by open markets on environmental legislation. The US unsuccessfully voted for Canada to review the designation.[1] The submission illustrates how international concern over the survival

---

[1] 'NAFTA panel won't review Canada's polar bear policy', CBC News, 6 June 2014. Available at: http://www.cbc.ca/news/canada/manitoba/nafta-panel-won-t-review-canada-s-polar-bear-policy-1.2667925

of polar bears has increased in recent years, with particular focus on the impacts of climate change and hunting on Canadian polar bear populations. It also reminds us of the role of the non-human (or perhaps the more than human) in shaping Arctic discourses and practices, including region building and the positioning of the Arctic as a local/regional space but also one connected to the global. The highly iconic polar bear, whale and seal have had, and continue to have, an extraordinary power to bring to the fore how the Arctic and Antarctic are understood and engaged with.

In other cases, demands for action provoke inaction or even dismissal. Take the example of the oil sands in northern Alberta and the attitude of the Harper administration to those who would seek to stop Canada from exploiting further such resources. While located south of the Arctic Circle, but nonetheless in a part of northern Canada, the oil sands have proven deeply controversial as critics complain that the exploitation of this resource is energy intensive and environmentally destructive. A visit by Hollywood actor Leonardo DiCaprio – like Emma Thompson part of what we might term a celebrity geopolitics issuing its own set of demands – in August 2014 to Fort McMurray, near oil sands operations, ensured that global media attention was higher on that particular area of northern Canada. DiCaprio has also been involved in a series of Green World Rising videos entitled 'Carbon', which have urged oil and gas resources to remain in the ground in order to tackle further climate warming. However, others did not welcome this critical attention on oil sands, as the Canadian Association of Petroleum Producers noted, 'Like Canadians, we [the industry] are growing tired of the fad of celebrity environmentalists coming into the region for a few hours or a few days, and offering their ideas and solutions to developing this resource.' The Energy Minister for Alberta, Diana McQueen, also remarked that, 'Aside from the economic opportunities the oil sands create for Albertans and Canadians, the taxes and royalties generated from the oil sands development provide funding for the infrastructure and programs that contribute to Canada's high standard of living.'

The demand for action on oil sands exploitation was rejected despite a celebrity intervention. More generally, such calls sit awkwardly with the Harper government's demands that Canada be considered as an 'emerging energy superpower' while at the same time expressing a reluctance to do anything that might harm the Canadian mining economies, including imposing tougher regulations on production standards and carbon emission reduction. If anything, there has been an intensification of interest in ensuring that Canada's existing and future resources are mapped, exploited, transported and administered. As Natural Resources Canada notes on its website:

Canada's Arctic is one of Earth's last frontiers for natural resource development. As a result mining and oil and gas development will be a key economic development instrument. The region is rich in diamonds, gold, oil and gas, base metals and iron ore. However, much of Canada's North has not been studied to a sufficient level to encourage and sustain resource investment and to inform land-use decisions such as the creation of parks and other protected areas. Recent and on-going activities, primarily as part of the Earth Sciences Sectors' Geo-Mapping for Energy and Minerals Program, are improving our knowledge of Canada's North through the acquisition and rapid release of new geoscience information for targeted areas with high potential for base metals (copper, nickel, iron, zinc and lead), precious metals (gold, silver, platinum), diamonds, and multiple commodities including rare metals. Similar work in the energy domain is providing the detailed geological framework needed to evaluate energy resource potential; supported by maps that indicate the major geological factors controlling the distribution of potential hydrocarbon resources.[2]

This extract shows that there is another kind of demanding at play. There is an unequivocal demand for further studies, especially in the field of mapping and resource evaluation, in order to enable a more robust appraisal of energy resource potential in the Canadian North rather than, say, a demand to leave those resources in place. Demands for certain kinds of action on behalf of campaigning groups sit uneasily with demands for other kinds of action, namely greater mapping and exploitation.

Our second theme is about how a demanding polar geopolitics makes and remakes understandings of proximity and distance. Take as an example the repeated attempts by Australia and New Zealand to establish marine protected areas (MPAs) in the Ross Sea and other areas of the Southern Ocean due south of the Australian coastline. The MPAs are well-established conservation management tools designed to protect the marine environment by restricting human activities such as fishing. By 2014, around 6500 MPAs were established covering some 2 per cent of the world's oceans; restrictions might include a so-called 'no-take' arrangement, which means that living resource exploitation is prohibited. The New Zealand government, with the support of other states and organizations such as the United

---

[2] Government of Canada, *National Resources Canada*. Available at http://www.nrcan.gc.ca/home

States and European Union, proposed a Ross Sea MPA in 2014–15 encompassing some 1.32 million square kilometres, with 1.25 million square kilometres being defined as 'no-take'. The argument for establishing the MPA was that the Ross Sea is one of the world's last areas largely untouched by human activities such as fishing. It is also an important ecosystem for whales, penguins, seals and finfish, including the commercially lucrative Patagonian Toothfish.

A marine survey, the Census of Antarctic Marine Life (CMAL), carried out in the midst of the 2007–8 International Polar Year, revealed that the Ross Sea was one of the world's least disrupted marine ecosystems. But with the development of commercial fisheries in the late 1990s led by proximate countries such as New Zealand, and distant water fishing fleets associated with Russia, South Korea and Spain, fears were expressed that the Ross Sea demanded greater environmental protection and resource restriction. New Zealand, the sponsor of the MPA for the Ross Sea, is the main commercial actor in the Ross Sea fishing economy.

What made the controversy over the MPA for the Ross Sea and associated demands for action interesting was that New Zealand is a claimant state and its claim to the Ross Dependency includes the Ross Sea. So while the Antarctic Treaty and its provisions demand that signatories place the contested sovereignty of Antarctica in abeyance for the purpose of international collaboration, the proposed Ross Sea MPA was never going to be straightforward. How would others, some of whom were quite distant to the Ross Sea, respond to the main commercial-political actor being a champion for resource conservation in the same region? Complicating matters still further, the proposal for the MPA had to be agreed upon by all the members of CCAMLR in order to be implemented. Under the terms of the convention, Article II seeks to reconcile what is described as 'rational use' of resources with an ecosystem-based management approach, which accounts not only for the impact of fishing on a particular species (e.g. Patagonian Toothfish) but also the likely impact on other elements of that ecosystem. In order to come to such judgements about appropriate conservation measures, and what is termed total allowable catch, reliable data on fish stock levels and distribution is vital, but modelling the Ross Sea ecosystem is far from an exact science.

In October 2014, it was announced that the CCAMLR parties had failed yet again to secure consensus on the Ross Sea MPA. The championing of these proposals by regional actors such as New Zealand, Australia and France (who supported other areas of the East Antarctic maritime region for MPA status) was not sufficient to generate the necessary consensus. While one extraterritorial party, the United States, supported the Ross Sea MPA proposal, it appeared that two others (China and Russia) were accused of deliberately wrecking the proposals at the meeting. While environmental campaigners were hoping that the unique and largely untouched ecosystem of the Ross Sea would elicit approval for environmental protection, both China and Russia were accused of pursuing their own commercial and geopolitical agendas. In other words, they were reluctant to approve restrictions on their own activities and suspicious that a claimant state such as New Zealand was eager to pursue a 'sovereignty agenda' in the name of environmental stewardship. The controversy over the MPA proposals, however, reflects a broader tension within CCAMLR, which struggles to reconcile 'rational use' of living resources on the one hand and conservation measures on the other.

As part of the fallout, moreover, opinion was divided as to the role that large, extraterritorial states such as China and Russia play in terms of demanding that their interests are heard. In Australia, there has been a long-standing unease at the role that countries like China are playing in the Antarctic, with suspicions being raised that Chinese investment in science and infrastructure is driven by a fundamental interest in resource exploitation and evaluation. As claimant states such as Australia and New Zealand understand, China and Russia do not recognize any Antarctic sovereignty claims and Russia continues to reserve the right to make a territorial claim in the future, while China considers the Antarctic and Southern Ocean to be a global common.

The role of 'distant' countries like China is also felt in the Arctic region where, as we noted in chapter 6, a form of polar Orientalism often prevails. One manifestation of such is a fear and mistrust of what 'Asian' actors like China, South Korea and India *really* want from the Polar Regions. So every investment, decision and action is scrutinized carefully by more locally 'rooted' actors, whether it be Canada, Norway, Australia or Denmark/Greenland. Not so long ago, the establishment of a Chinese Embassy in Iceland, for example, unleashed press speculation that the Chinese were obviously going to demand more from the Arctic because their building was considered to be so large for a diplomatic mission to such a small country. In Svalbard, speculation was also rife that a Chinese investor was eager to purchase a 200 km plot of land in the Norwegian archipelago. Some Norwegian commentators feared that such a sale would enable China to have a 'foothold' in the Arctic, and thus facilitate a greater physical presence beyond their existing research station in Svalbard.

Our third theme regarding demands revolves around responsibilities. How do demands for action and inaction generate responsibilities to pursue things like access, solidarity, belonging, redress and justice? While we do not have the space to interrogate all these things, we might consider how a demanding polar geopolitics is one where human agents are urged to act and even intervene. Some indigenous leaders, for example, have articulated climate change as a human rights issue and strive to have their voices heard on this in international policymaking circles. Inuit leaders and community representatives in Canada have argued that cultural survival is dependent on the continued presence of ice and snow. In 2005, Sheila Watt-Cloutier, the former international chair of the Inuit Circumpolar Council (then still known as the Inuit Circumpolar Conference), submitted a 167-page petition to the Inter-American Commission on Human Rights on behalf of all Inuit of the Arctic regions of the United States and Canada. The petition dealt specifically with an argument that the violation of Inuit human rights was caused by greenhouse gas emissions from the United States. In it, Watt-Cloutier argued that climate change is harming every aspect of Inuit life and culture and drew attention to the intimate relations between Inuit and the Arctic environment:

> Like many indigenous peoples, the Inuit are the product of the physical environment in which they live. The Inuit have fine-tuned tools, techniques and knowledge over thousands of years to adapt to the arctic environment. They have developed an intimate relationship to their surroundings, using their understandings of the arctic environment to develop a complex culture that has enabled them to survive on scarce resources. The culture, economy and identity of the Inuit as an indigenous people depend upon the ice and snow. (Watt-Cloutier 2005: 1)

The petition goes on to describe how this delicate balance between Inuit and the environment is now threatened by climate change and how Inuit are struggling to adapt to the transformations brought about by global warming. A careful reading of the petition, however, reveals that it is about more than just Inuit concerns with the impacts of climate change. Its recasting of climate change as a human rights issue, not just an environmental one, draws attention to the position of Inuit as indigenous people within nation states, and in particular to broader aspects of indigenous rights and to the responsibilities of governments to address their concerns and recognize those rights. But there is a regional texture to climate change. In the far north of Greenland, for instance, while hunters express concern over the thinning of sea ice, poor dog sledging conditions and the inability to harvest seals, local farmers in the sheep farming districts of the south coast express amazement at bumper harvests of potatoes and the growth of succulent cucumbers. In both areas, climate change is having noticeable impacts. While hunting and fishing households in small, remote northern communities are facing an uncertain future and are encountering difficulties in accessing traditional hunting and fishing areas, sheep farming households in equally remote southerly districts are taking advantage of a warmer climate that may allow them additional income through greater yields of vegetables and increasing areas of fertile pasture. Climate change may be an issue of cultural survival for some, but for others it also holds the promise of economic prosperity (Nuttall 2009).

Understanding climate change and its impacts and consequences for indigenous people and other Arctic residents demands from us a particular attention to how we think of vulnerability and resilience and how we approach sensitive and politically difficult issues of adaptation and environmental governance. Climate change is but one of several often interrelated circumstances affecting livelihoods throughout the circumpolar North. Indigenous peoples in particular are exposed to a variety of different drivers of change alongside societal and environmental transformation that can increase their exposure to global processes and often affect their abilities to respond creatively and effectively. Inuit can no longer adapt, relocate or change resource use activities as easily as they may have been able to in the past, for instance, because most now live in permanent communities, in greatly circumscribed social and economic situations, where their hunting and fishing activities are determined to a large extent by resource management regimes, quota systems and political decision-making bodies, which are sometimes far removed from their communities and local and global markets (Nuttall et al. 2005). In northern Greenland, for example, while the changing nature of sea ice has had a dramatic effect on fishing and hunting activities, quota systems for narwhals and Greenland halibut often have considerably more impact on livelihoods than a shifting biological environment. Local and regional systems of management of living resources are often implemented by government departments and institutions, in accordance with the guidance and recommendations of international organizations, that deal with the management of fisheries and conservation of marine mammals. Research on vulnerability and resilience, including Arctic Council projects such as the Arctic Resilience Report, fail to acknowledge and explicitly deal with such management decisions and the

184

local-regional-global nature of the institutional cultures of international organizations, as well as the impacts and legacies of colonialism, the relocation of people from hunting and reindeer camps and small villages into urban centres, the long-term and cyclical effects of residential schools, and rapid social change for indigenous peoples. It may well be that the long-lasting impact of environmental campaigning against seal hunting and the resulting trade bans on products from the hunt have far more significance for some Inuit hunters than climate change and melting ice. Elsewhere, livelihoods are challenged by governance systems and institutions that often inhibit and constrain locally specific long-term resource availability, and the rights of individuals and communities to access those resources (e.g. Heikkinen et al. 2011).

As the Iñupiat of Shishmaref are discovering, the mobility their ancestors once used to respond to shifts in the pattern and state of their resource base is no longer an option for contemporary communities. In contemplating the severity of their present situation and the urgency to move to surer ground, the community is working through the Shishmaref Erosion and Relocation Committee to relocate to the site of Tin Creek under a project jointly coordinated by local, state and federal authorities. Across much of the Inuit world, people have been moved by the state into communities which are dependent on government infrastructure, subsidies and other investment. Inuit in Nunavut, for example, live in communities resulting directly from Canadian government policy, implemented in the 1950s and 1960s, of moving them from nomadic camps into permanent settlements. In today's social, political and economic climate, movement and seasonal migration to remain in contact with animals and, more broadly, to maintain traditional Inuit hunting livelihoods would seem to be virtually impossible without some form of income from part-time, seasonal or permanent employment, or government assistance. The effect is to limit and constrain traditional hunting and fishing activities. Wildlife management and conservation, aimed in principle at protecting and conserving wildlife, also often restricts access to resources. Combined with ecosystem shifts, this magnifies the potential effects of global climate change on Inuit communities.

Two examples might illuminate this point more. First, we might think about how indigenous peoples in the Arctic have demanded that national governments take responsibility for their interventions in the past. In Canada, the so-called High Arctic Relocation remains a highly emotive and controversial element of Canadian political history. In the early 1950s, at the height of the Cold War, 87 Inuit were relocated to the Canadian High North. While the exercise was conceived of in part as a humanitarian operation designed to improve the lives of those families, it was widely suspected that the relocation was driven by a geopolitical agenda, in which Canada was eager to assert its sovereignty through 'effective occupation'. In other words, the bodies of Inuit were invested with considerable sovereign power, albeit one directed and dictated by the Canadian government.

The end result of the relocation was to move these families from northern Quebec to Ellesmere Island and Cornwallis Island. The idea was that local community members would be drafted in to help the new families learn a subsistence lifestyle suitable in the High Arctic, and contribute in turn to a reduction in welfare dependency. By way of contrast, Inuit and academic scholars contend that the relocation was a biopolitical experiment, an attempt to manage a 'population' and to settle territories insufficiently protected by those who might challenge Canadian sovereignty, both the United States and the Soviet Union. The families in question were left with insufficient provisions and support and earlier guarantees to offer them resettlement back in Quebec if they wished were not honoured.

Some 20 years later, Inuit, including those who were part of those relocated families, began a compensation campaign, and demanded that the Canadian government recognize the hardship it had asked the families to endure. They also charged governments with perpetuating a mendacious campaign of disinformation. In the 1990s a Royal Commission on Aboriginal Peoples heard evidence about the relocation initiative and concluded that the government should apologize and compensate the surviving members of those families. A trust fund was eventually established in 1996 but the demanded apology was not forthcoming. In August 2010, the Minister of Indian Affairs and Northern Development, John Duncan, finally delivered an official apology in August 2010:

> The Government of Canada deeply regrets the mistakes and broken promises of this dark chapter of our history and apologizes for the High Arctic Relocation having taken place. We would like to pay tribute to the relocatees for their perseverance and courage ... The relocation of Inuit families to the High Arctic is a tragic chapter in Canada's history that we should not forget, but that we must acknowledge, learn from and teach our children. Acknowledging our shared history allows us to move forward in partnership and in a spirit of reconciliation.

The point for us about the tragic story of the High Arctic Relocation of seven families from northern Quebec is to remind us that polar geopolitics is demanding in terms of responsibilities for the past as well as the present. Contemporary Canadian policies and strategies towards the Arctic are inflected and informed by how past encounters haunt the present and the demands that shape how governments act and react to indigenous communities, and their calls for autonomy, justice and resource sharing.

Future Polar Demands

Another way of thinking about demands and responsibilities might be to consider what we do and do not know about the Polar Regions, and the pressures that are placed upon us as a consequence. The relationship between knowledge and ignorance is never politically innocent. Our point here is to think about how polar geopolitics generates demands for knowledge. Coastal states such as Canada and Australia urge their scientists to improve the mapping and surveying of extended continental shelves. Commercial actors such as the shipping industry demand greater knowledge of sea ice distribution and thickness, while polar scientists warn that satellites struggle to reconcile accurate measurement with a realization that physical phenomena such as pools of melted water on the surface of ice confuse onboard instrumentation. It is sobering to hear and see presentations from representatives of shipping companies positing caution and doubt about the commercial opportunities posed by Arctic shipping routes, and expressing scepticism that diminishing sea ice automatically means that accessibility for ships improves. Forecasting remains a fraught business and scientists have warned that satellite monitoring is no panacea for surface level observation and measurement.

The role of knowledge (and by association ignorance) has been critical in mediating polar geopolitics and the demands that it might make on people, places and objects. Physical and environmental research into the Arctic and Antarctic contributes, we would argue, to both an ecological fragility and economic opportunity discourse. On the one hand, research into snow, ice and climate change emphasizes in the main a more fragile view of polar environments, where melting and a sense of profound perturbation predominate as opposed to a sense of polar environments as robust, unchanging and hostile. On the other hand, geosciences such as geology and oceanography have been important accomplices in state-sanctioned projects, which demand greater knowledge of terrestrial and maritime environments. While the distinction can be overdrawn, it makes us wonder about the demands that are made on academics and other knowledge producers, including indigenous parties, in this contemporary era.

Our book has touched upon what we consider to be some of the most important dimensions of the contemporary geopolitics of the Polar Regions. There will be considerably more work to be undertaken as we and other scholars demand more critical forms of polar geopolitics (Powell and Dodds 2014). In other words, we need to challenge those forms of polar geopolitics that reproduce uncritical 'scramble' discourses, while resisting the temptation to exaggerate, to simplify and to marginalize. In order to do that, there is a need to demand more attention to historical connections and contexts, including colonialism and imperialism, which ensured that the Arctic and Antarctic were intimately tied to colonial-era science, commerce and geopolitics. These regions were not 'poles apart'. In addition, the contemporary manifestations of polar geopolitics need to be better understood, albeit in a context where academic knowledge and other forms of expertise are readily accessed and consumed by stakeholders including states, corporations, pressure groups and campaigning organizations. Our commitment to developing a critical polar geopolitics is demanding and will continue to be so in the future.

We do not expect the demands placed on the Polar Regions to diminish, if anything they will intensify. It will take immense collective self-restraint to imagine the Arctic and Antarctic not being exploited further. In the Arctic those resource pressures will come into contact with indigenous and northern communities demanding that their wishes and interests are heard and respected, while international agreements and demands for conservation and environmental protection will be tested in Antarctica. The Polar Regions have witnessed firsthand humankind's capacity to act as a geophysical force, as we continue to inhabit an increasingly hot planet with all the demands that an ever-growing population bring with it. Whatever happens, the fate for the Polar Regions will be increasingly slippery, as we face the future possibility of a world without ice.

REFERENCES

ACIA (2004) *Impacts of a Warming Arctic: Arctic Climate Impact Assessment.* Cambridge: Cambridge University Press.

ACIA (2005) *Arctic Climate Impact Assessment: The Scientific Report.* Cambridge: Cambridge University Press.

Agnew, John A. (1998) *Geopolitics: Re-Visioning World Politics.* London: Routledge.

AMAP (2011) *Changes in Arctic Snow, Water, Ice and Permafrost.* Oslo: Arctic Monitoring and Assessment Programme.

AMAP (2014) *Arctic Ocean Acidification Overview Report.* Oslo: Arctic Monitoring and Assessment Programme.

Anderson, Alun (2009) *After the Ice: Life, Death, and Geopolitics in the New Arctic.* New York: Smithsonian Books.

Anderson, David G. (2000) *Identity and Ecology in Arctic Siberia: the Number One Reindeer Brigade.* Oxford: Oxford University Press.

Anderson, David G. and Mark Nuttall (2004) *Cultivating Arctic Landscapes: Knowing and Managing Animals in the Circumpolar North.* Oxford: Berghahn Books.

Anderson, Mitch, Matt Finer, Daniel Herriges, Andrew Miller and Atossa Soltani (2009) *ConocoPhillips in the Peruvian Amazon.* A report by Amazon Watch and Save America's Forests. Available at: http://saveamericasforests.org/Yasuni/Publications/conoco2009.pdf

Arctic Council (2009) *Arctic Marine Shipping Assessment.* Tromsø: Protection of the Arctic Marine Environment (PAME).

Arctic Economic Council (2014) Fostering Circumpolar Business Partnerships. Available at: http://www.arctic-council.org/index.php/en/arctic-economic-council

189

Atwood, Margaret (1991) *Strange Things*. Oxford: Oxford University Press.

Banerjee, Subhankar (2003) *Arctic National Wildlife Refuge: Seasons of Life and Land*. Seattle, WA: Mountaineers Books.

Barbier, Edward B. (2011) *Scarcity and Frontiers: How Economies Have Developed Through Natural Resource Exploitation*. Cambridge: Cambridge University Press.

Barume, Albert K. (2010) *Land Rights of Indigenous Peoples in Africa*. Copenhagen: IWGIA.

Berger, Thomas R. (1977) *Northern Frontier, Northern Homeland: the Report of the Mackenzie Valley Pipeline Inquiry*. Ottawa: Minister of Supply and Services Canada.

Bergin, Anthony and Marcus Haward (2007) 'Frozen Assets: Securing Australia's Antarctic Future'. Strategic Insight 34. Australian Strategic Policy Institute.

Berland, Jody (2009) *North of Empire: Essays on the Cultural Technologies of Space*. Durham, NC: Duke University Press.

Bert, Melissa (2012) 'The Arctic Is Now: Economic and National Security in the Last Frontier'. *American Foreign Policy Interests: The Journal of the National Committee on American Foreign Policy* 34 (1): 5–19.

Berton, Pierre (1972) *The Impossible Railway: The Building of the Canadian Pacific*. New York: Alfred A. Knopf.

Billig, Michael (1995) *Banal Nationalism*. London: Sage.

Bird, Kenneth J., Ronald R. Charpentier, Donald L. Gautier, David W. Houseknecht, Timothy R. Klett, Janet K. Pitman, Thomas E. Moore, Christopher J. Schenk, Marilyn E. Tennyson, and Craig J. Wandrey (2008) Circum-Arctic Resource Appraisal: Estimates of Undiscovered Oil and Gas North of the Arctic Circle, US Geological Survey Fact Sheet 2008-3049.

Blank, Stephen (2013) 'China's Arctic Strategy'. *The Diplomat*, 20 June. Available at: http://thediplomat.com/2013/06/chinas-arctic-strategy/

Bloom, Lisa (1993) *Gender on the Ice*. Minneapolis, MN: University of Minnesota Press.

Bloom, Lisa, Elena Glasberg and Laura Kay (2008) 'Introduction: New Poles, Old Imperialism?'. *Scholar and Feminist Online* 7(1). Available at: http://sfonline.barnard.edu/ice/intro_01.htm

Bocking, Stephen (2009) 'A Disciplined Geography: Aviation, Science, and the Cold War in Northern Canada, 1945–1960'. *Technology and Culture* 50: 265–90.

190

Bockstoce, John R. (2009) *Furs and Frontiers in the Far North: The Contest among Native and Foreign Nations for the Bering Strait Fur Trade*. New Haven, CT: Yale University Press.

Bone, Robert M. (2003) *The Geography of the Canadian North: Issues and Challenges*. Oxford: Oxford University Press.

Bone, Robert M. (2009) *The Canadian North: Issues and Challenges*. New York: Oxford University Press.

Brady, Anne-Marie (2012) 'Polar Stakes: China's Polar Activities as a Benchmark for Intentions'. *China Brief* 12(14), 19 July.

Bravo, Michael T. (2009) 'Voices from the Sea Ice and the Reception of Climate Impact Narratives'. *Journal of Historical Geography* 35: 256–78.

Bravo, Michael T. and Gareth Rees (2006) 'Cyro-Politics: Environmental Security and the Future of Arctic Navigation'. *Brown Journal of World Affairs* 13: 205–16.

Brekke, Harald (2014) 'Defining and Recognizing the Outer Limits of the Outer Continental Shelf in the Polar Regions'. In Richard C. Powell and Klaus Dodds (eds) *Polar Geopolitics*. Cheltenham: Edward Elgar, pp. 38–54.

Bridge, Gavin (2013) 'Territory now in 3D!' *Political Geography* 34 (3): 55–7.

Brøsted, Jens and Mads Fægteborg (1995) 'Thules invasion og befolkningens fordrivelse'. *Nordisk Kontakt* 10: 15–21.

Byers, Michael (2013) *International Law and the Arctic*. Cambridge: Cambridge University Press.

Cameron, Emilie (2008) 'Cultural Geographies Essay: Indigenous Spectrality and the Politics of Postcolonial Ghost Stories'. *Cultural Geographies* 15: 383–93.

Carmody, Pádraig (2011) *The New Scramble for Africa*. Cambridge: Polity.

Cater, Tara and Arn Keeling (2013) '"That's Where Our Future Came From": Mining, Landscape, and Memory in Rankin Inlet, Nunavut'. *Études/Inuit/Studies* 37(2): 59–82.

Caulfield, Richard A. (2000) 'The Political Economy of Renewable Resource Management in the Arctic'. In Mark Nuttall and Terry V. Callaghan (eds) *The Arctic: Environment, People, Policy*. Amsterdam: Harwood Academic, pp. 485–514.

Caulfield, Richard A. (2007) *Greenlanders, Whales and Whaling: Sustainability and Self-Determination in the Arctic*. Hanover, NH: University of New England Press.

CBC News (2011) '17th Century Chinese Coin Found in Yukon', CBC News, 1 November. Available at: http://www.cbc.ca/news/canada/north/story/2011/11/01/north-acient-coin-found.html

Chapin, Stuart F., Mathew Berman, Terry V. Callaghan, Peter Convey, Anne-Sophie Crépin, Kjell Danell, Hugh Ducklow, Bruce Forbes, G. Kofinas, Anthony David McGuire, Mark Nuttall, Ross Virginia, Oran Young and Sergei A. Zimov (2005) 'Polar Systems'. In Millennium Ecosystem Assessment, *Ecosystems and Human Well-Being: Current State and Trends*. Washington, DC: Island Press, pp. 717–43.

Chaturvedi, Sanjay (2005) 'Arctic Geopolitics Then and Now'. In Mark Nuttall (ed.) *Encyclopedia of the Arctic*. London: Routledge, pp. 724–30.

Cherry-Garrard, Apsley (1922) *The Worst Journey in the World*. New York: Carroll & Graf.

China Daily (2012) 'China Has a Role in Safeguarding the Arctic'. *China Daily*, 29 June.

Chown, S.L., J.E. Lee, K.A. Hughes, J. Barnes, P.J. Barrett, D.M. Bergstrom, P. Convey, et al. (2012) 'Challenges to the Future Conservation of the Antarctic'. *Science* 337(6091), 158–9.

Churchill, Ward (2002) *Struggle for the Land: Native North American Resistance to Genocide, Ecocide and Colonization*. San Francisco, CA: City Lights Books.

Cloud, John (2002) 'American Cartographic Transformations during the Cold War'. *Cartography and Geographic Information Science* 29: 261–82.

Cohen, Stanley (1972) *Folk Devils and Moral Panics*. London: MacGibbon & Kee.

Collis, Christy (2009) 'The Australian Antarctic Territory: A Man's World?'. *Signs* 34: 514–19.

Collis, Christy and Klaus Dodds (2008) 'Assault on the Unknown: The Historical and Political Geographies of the International Geophysical Year (1957–8)'. *Journal of Historical Geography* 34: 555–73.

Colpært, Alfred (2006) 'The Forgotten Uranium Mine of Paukkajanvaara, North Karelia, Finland'. *Nordia Geographical Publications* 35(2): 31–8.

Cornell, Sarah, Bruce Forbes, Donald McLennan, Ulf Molau, Mark Nuttall, Paul Overduin and Paul Wassman (2013) 'Thresholds in the Arctic'. In Arctic Council, *Arctic Resilience Interim Report 2013*. Stockholm: Stockholm Environment Institute/Arctic Council, pp. 35–67.

Cosgrove, Denis E. (2001) *Apollo's Eye*. Baltimore, MD: Johns Hopkins University Press.

Costanza, Robert, Fransisco Andrade, Paula Antunes, Marjan van den Belt, Dee Boersma, Don Boesch et al. (1998) 'Principles for Sustainable Governance of the Oceans'. *Science* 281: 198–9.

Craciun, Adriana (2009) 'The Scramble for the Arctic'. *Interventions: International Journal of Postcolonial Studies* 11(1): 103–14.

Cracuin, Adriana (2012) 'The Franklin mystery'. *Literary Review of Canada*, May. Available at: http://reviewcanada.ca/magazine/2012/05/the-franklin-mystery/

Cracuin, Adriana (2014) 'The Franklin Relics in the Arctic Archive'. *Victorian Literature and Culture* 42 (1): 1–31.

Crampton, Jeremy (2010) 'Cartographic Calculations of Territory'. *Progress in Human Geography* 35(1): 92–103.

Crawford, Amy (2013) 'When an Iceberg Melts, Who Owns the Riches Beneath the Ocean?' *Smithsonian Magazine*, April. Available at: http://www.smithsonianmag.com/innovation/when-an-iceberg-melts-who-owns-the-riches-beneath-the-ocean-4800218/?no-ist

Cresswell, Tim (2006) *On the Move: Mobility in the Modern Western World*. London: Taylor & Francis.

Cussler, Clive (2008) *Arctic Drift*. New York: G.P. Putnam's Sons.

Dahl, Jens (1993) 'Indigenous Peoples of the Arctic'. In Charlotta Friborg and Svenolof Karlsson (eds) *Arctic Challenges*. Stockholm: The Nordic Council, pp. 103–30.

Dahl, Jens (2000) *Saqqaq: An Inuit Hunting Community in the Modern World*. Toronto: University of Toronto Press.

Dalby, Simon (1991) 'Critical Geopolitics: Discourse, Difference and Dissent'. *Environment and Planning D: Society and Space* 9(3): 261–83.

Dalby, Simon (2009) *Security and Environmental Change*. Cambridge: Polity.

Darby, Andrew (2014) 'China's Antarctica Satellite Base Plans Spark Concerns'. *The Sydney Morning Herald*, 12 November. Available at: http://www.smh.com.au/world/chinas-antarctica-satellite-base-plans-spark-concerns-20141112-11l3wx.html

David, Rob (2000) *The Arctic in the British Imagination, 1818–1914*. Manchester: Manchester University Press.

Davis, William Morris (1910) 'Antarctic Geology and Polar Climates'. *Proceedings of the American Philosophical Society* 49(195): 200–2.

Dean, Katrina, Simon Naylor, Simone Turchetti and Martin Siegert (2008) 'Data in Antarctic Science and Politics'. *Social Studies of Science* 38: 571–604.

Dennis, Michael A. (2003) 'Postscript: Earthly Matters: On the Cold War and the Earth Sciences'. *Social Studies of Science* 33: 809–19.

Descola, Philippe (1994) *In the Society of Nature: A Native Ecology in Amazonia*. Cambridge: Cambridge University Press.

Deutsche Welle (2013) 'All eyes on the Arctic'. Deutsche Welle, 16 May. Available at: http://www.dw.de/all-eyes-on-the-arctic-council/a-16811193

Dittmer, Jason, Sami Moisio, Alan Ingram and Klaus Dodds (2011) 'Have You Heard the One About the Disappearing Ice? Recasting Arctic Geopolitics'. *Political Geography*, 30(4), 202–14.

Dodds, Klaus (2002) *Pink Ice: Britain and the South Atlantic Empire*. London: I.B. Tauris.

Dodds, Klaus (2010) 'Flag Planting and Finger Pointing: The Law of the Sea, the Arctic and the Political Geographies of the Outer Continental Shelf'. *Political Geography* 29: 63–73.

Dodds, Klaus (2012) *The Antarctic: A Very Short Introduction*. Oxford: Oxford University Press.

Dodds, Klaus (2013) 'Anticipating the Arctic and the Arctic Council: Pre-Emption, Precaution and Preparedness'. *Polar Record* 49(2): 193–203.

Dodds, Klaus and David Atkinson (eds) (2000) *Geopolitical Traditions: A Century of Geopolitical Thought*. New York: Routledge.

Dodds, Klaus and Alan D. Hemmings (2009) 'Frontier Vigilantism? Australia and Contemporary Representations of Australian Antarctic Territory'. *Australian Journal of Politics & History* 55(4): 513–29.

Doel, Ronald E. (2003) 'Constituting the Postwar Earth Sciences: The Military's Influence on the Environmental Sciences in the USA after 1945'. *Social Studies of Science* 33(5): 635–66.

Doel, Ronald E., Tanya J. Levin and Mason K. Marker (2006) 'Extending Modern Cartography to the Ocean Depths: Military Patronage, Cold War Priorities, and the Heezen-Tharp Mapping Project 1952–1959'. *Journal of Historical Geography* 32: 605–26.

Doel, Ronald E., Urban Wråkberg and Suzanne Zeller (2014) 'Science, Environment, and the New Arctic'. *Journal of Historical Geography* 44: 2–14.

Driver, Felix (2000) *Militant Geography*. Oxford: Blackwell.

Dunaway, Finis (2006) 'Reframing the Last Frontier: Subhankar Banerjee and the Visual Politics of the Arctic National Wildlife Refuge'. *American Quarterly* 58: 159–80.

Ebinger, Charles K. and Evie Zambetakis (2009) 'The Geopolitics of Arctic Melt'. *International Affairs* 85(6): 1215–32.

Edwards, Robert (2006) *White Death: Russia's War on Finland 1939–40*. London: Weidenfeld & Nicolson.

Elden, Stuart (2013) 'Securing the Volume: Vertical Geopolitics and the Depth of Power'. *Political Geography* 34: 35–51.

Elliott, Lorraine M. (1994) *International Environmental Politics: Protecting the Antarctic*. London: Macmillan.

Emmerson, Charles (2010) *The Future History of the Arctic*. New York: PublicAffairs.

English, John (2013) *Ice and Water: Politics, Peoples, and the Arctic Council*. London: Palgrave.

Exner-Pirot, Heather (2013) 'What Is the Arctic a Case Of? The Arctic as a Regional Environmental Security Complex and the Implications for Policy'. *The Polar Journal* 3(1): 120–35.

Farish, Matthew (2010) *The Contours of America's Cold War*. Minneapolis, MN: University of Minnesota Press.

Findlay, Ronald and Mats Lundahl (1999) Resource-Led Growth – A Long-Term Perspective: The Relevance of the 1870–1914 Experience for Today's Developing Economies. UNU Working Papers No. 162, Helsinki: UNU World Institute for Development Economics Research.

Fitzhugh, William W. and Valerie Chaussonnet (1994) *Anthropology of the North Pacific Rim*. Washington, DC: Smithsonian Institution Press.

Fjellheim, Rune S. and John B. Henriksen (2006) 'Oil and Gas Exploitation on Arctic Indigenous Peoples' Territories: Human Rights, International Law, and Corporate Social Responsibility'. *Journal of Indigenous Peoples' Rights* 4: 8–52.

Flake, Lincoln E. (2013) 'Navigating an Ice-Free Arctic: Russia's Policy on the Northern Sea Route in an Era of Climate Change'. *Rusi Journal* 158(3), 44–52.

Fondahl, Gail and Stephanie Irlbacher-Fox (2009) *Indigenous Governance in the Arctic: A Report for the Arctic Governance Project*. Prepared for the Walter and Duncan Gordon Foundation. Available at: http://www.arcticgovernance.org/compendium.137 742.en.html

Fondahl, Gail and Anna Sirina (2006) 'Oil Pipeline Development and Indigenous Rights in Eastern Siberia'. *Indigenous Affairs* 2–3/06: 58–67.

Forsyth, James (1992) *A History of the Peoples of Siberia: Russia's North Asian Colony 1581–1990*. Cambridge: Cambridge University Press.

Frayling, Christopher (2014) *The Yellow Peril: Dr. Fu Manchu and the Rise of Chinaphobia*. London: Thames and Hudson.

Gerhardt, Hannes, Philip E. Steinberg, Jeremy Tasch, Sandra J. Fabiano and Rob Shields (2010) 'Contested Sovereignty in a Changing Arctic'. *Annals of the Association of American Geographers* 100: 992–1002.

Gill, Duane A. and J. Steven Picou (1997) 'The Day the Water Died: Cultural Impacts of the *Exxon Valdez* Oil Spill'. In J. Steven Picou, Duane A. Gill and Maurie J. Cohen (eds) *The Exxon Valdez Disaster: Readings on a Modern Social Problem*. Dubuque, IA: Kendall/Hunt Publishing Co., pp. 167–84.

Glasberg, Elena (2013) *Antarctica as Cultural Critique*. Basingstoke: Macmillan.

Gorbachev, Mikhail (1987) Speech in Murmansk at the Ceremonial Meeting on the Occasion of the Presentation of the Order of Lenin and the Gold Star to the City of Murmansk. Available at: http://www.arctic.or.kr/files/pdf/m2/m22/1/m22_1_eng.pdf

Gordon, Avery (1997) *Ghostly Matters*. Minneapolis, MN: University of Minnesota Press.

Grace, Sherrill E. (2001) *Canada and the Idea of North*. Montreal: McGill-Queens University Press.

Graham, Stephen (2004) 'Vertical Geopolitics: Baghdad and After'. *Antipode* 36(1): 12–23.

Gregory, Derek (1994) *Geographical Imaginations*. Oxford: Blackwell.

Gregory, Derek (2004) *The Colonial Present*. Malden, MA: Blackwell.

Griffin, Duan (2004) 'Hollow and Habitable Within: Symmes's Theory of Earth's Internal Structure and Polar Geography'. *Physical Geography* 25: 382–97.

Griffith, Tom (2007) *Silencing the Silence: Voyaging to Antarctica*. Cambridge, MA: Harvard University Press.

Griffiths, Franklin (1999) 'The Northwest Passage in Transit'. *International Journal* 54(2): 189–202.

Hamblin, Jacob (2005) *Oceanographers and the Cold War*. Seattle, WA: University of Washington Press.

Hamilton, Clive (2010) *Requiem for a Species*. London: Earthscan.

Hannah, Matthew G. (2010) '(Mis)adventures in Rumsfeld Space'. *GeoJournal* 75: 397–406.

Harper, Christine (2008) *Weather by the Numbers: The Genesis of Modern Meteorology*. Cambridge, MA: MIT Press.

Hastrup, Kirsten (2009) 'The Nomadic Landscape: People in a Changing Arctic Environment'. *Geografisk Tidsskrift – Danish Journal of Geography* 109(2): 181–9.

Hastrup, Kirsten (2014) *Anthropology and Nature*. New York: Routledge.

Hayes, Isaac Israel (1868) 'The Progress of Arctic Discovery'. Lecture to the American Geographical and Statistical Society, New York, 12 November.

Heikkinen, Hannu, Simo Sarkki, Mikko Jokinen and David Fornander (2010) 'Global Area Conservation Ideals versus the Local Realities of Reindeer Herding in Northernmost Finland'. *International Journal of Business and Globalisation* 4(2): 110–30.

Heikkinen, Hannu, Outi Moilanen, Mark Nuttall and Simo Sarkki (2011) 'Managing Predators, Managing Reindeer: Contested Conceptions of Predator Policies in Finland's Southeast Reindeer Herding Area'. *Polar Record* 47(242): 218–30.

Heininen, Lassi (2005) 'Impacts of Globalization, and the Circumpolar North in World Politics'. *Polar Geography* 29(2): 91–102.

Heininen, Lassi and Chris Southcott (eds) (2010) *Globalization of the Circumpolar North*. Fairbanks, AK: University of Alaska Press.

Hemmings, Alan D. (2014) 'The Antarctic Treaty System [The Year in Review – 2012]'. *New Zealand Yearbook of International Law* 10: 237–43.

Hemmings, Alan D., Donald R. Rothwell and Karen N. Scott (eds) (2012) *Antarctic Security in the Twenty-First Century: Legal and Policy Perspectives*. Abingdon: Routledge.

Hill, Jen (2008) *The Arctic in the Nineteenth-Century British Imagination*. Albany, NY: State University of New York Press.

Hodgkins, Richard (2014) 'The Twenty-First-Century Arctic Environment: Accelerating Change in the Atmospheric, Oceanic and Terrestrial Spheres'. *The Geographical Journal* 180(4), 429–36.

Høeg, Peter (1993) *Miss Smilla's Feeling for Snow*. London: Harvill.

Holmes, James (2012) 'Open Seas: The Arctic is the Mediterranean of the 21st Century'. *Foreign Policy*, 29 October.

Howard, Roger (2009) *The Arctic Gold Rush: The New Race for Tomorrow's Natural Resources*. London: Bloomsbury Academic.

Howkins, Adrian (2008) 'Defending Polar Empire: Opposition to India's Proposal to Raise the "Antarctic Question" at the United Nations in 1956'. *Polar Record* 44: 35–44.

Howkins, Adrian (2011) 'Melting Empires? Climate Change and Politics in Antarctica since the International Geophysical Year'. *Osiris* 26: 180–97.

Humpert, Malte and Andreas Raspotnik (2012) 'From Great Wall to Great White North: Explaining China's Politics in the Arctic'.

Available at: http://www.thearcticinstitute.org/2012/08/from-great-wall-to-great-white-north.html

Huntington, Henry P. (1992) *Wildlife Management and Subsistence Hunting in Alaska*. Seattle, WA: University of Washington Press.

Huntington, Henry P. and Shari Fox (2005) 'The Changing Arctic: Indigenous Perspectives'. In ACIA (ed.) *Impacts of a Warming Climate: Arctic Climate Impact Assessment*. New York: Cambridge University Press, pp. 61–98.

Inter-American Court of Human Rights (2001) *Case of the Mayagna (Sumo) Awas Tingni Community v. Nicaragua*. Judgment of August 31, 2001. Available at: http://www.corteidh.or.cr/docs/casos/articulos/seriec_79_ing.pdf

IPCC (2013) *Climate Change 2013: The Physical Science Basis*. Bern: IPCC Working Group.

Irlbacher-Fox, Stephanie (2005) 'Land Claims'. In Mark Nuttall (ed.) *Encyclopedia of the Arctic*. New York: Routledge, pp. 1152–6.

Jakobson, Linda (2010) 'China Prepares for an Ice-Free Arctic'. SIPRI Insights on Peace and Security No. 2010/2 March.

Jakobson, Linda (2012) 'Northeast Asia turns its attention to the Arctic'. NBR Analysis Brief, 17 December.

Jentoft, Svein (2003) 'Introduction'. In Svein Jentoft, Henry Minde and Ragnar Nilsen (eds) *Indigenous Peoples: Resource Management and Global Rights*. Delft: Eburon Academic Publishers.

Josephson, Paul (2002) *Industrialized Nature: Brute Force Technology and the Transformation of the Natural World*. Washington, DC: Island Press.

Jurjevics, Juris (2005) *The Trudeau Vector*. New York: Viking Adult.

Kafarowski, Joanna (2009) 'Gender, Culture, and Contaminants in the North'. *Signs* 34(3): 494–9.

Kalland, Arne and Frank Sejersen (2005) *Marine Mammals and Northern Cultures*. Edmonton: Canadian Circumpolar Institute Press.

Kaplan, Robert D. (2010) 'The Geography of Chinese Power: How Far Can Beijing Reach on Land and at Sea?'. *Foreign Affairs*, May/June. Available at: http://www.foreignaffairs.com/articles/66205/robert-d-kaplan/the-geography-of-chinese-power

Keeling, Arn and John Sandlos (2009) 'Environmental Justice Goes Underground? Historical Notes from Canada's Northern Mining Frontier'. *Environmental Justice* 2(3): 117–25.

Keskitalo, E. Carina H. (2004) *Negotiating the Arctic: The Construction of an International Region*. New York: Routledge.

Kingdom of Denmark (2008) The Ilulissat Declaration. Ilulissat,

REFERENCES

Greenland, 27–29 May 2008. Available at: http://www.oceanlaw. org/downloads/arctic/Ilulissat_Declaration.pdf

Kinney, D.J. (2013) 'Selling Greenland: The Big Picture Television Series and the Army's Bid for Relevance during the Early Cold War'. *Centaurus* 55(3): 344–57.

Klare, Michael T. (2013) 'Rushing for the Arctic's Riches'. *New York Times*, 7 December. Available at: http://www.nytimes. com/2013/12/08/opinion/sunday/rushing-for-the-arctics-riches. html?pagewanted=all

Kollin, Susan (2000) 'The Wild, Wild North: Nature Writing, Nationalist Ecologies, and Alaska'. *American Literary History* 12: 41–78.

Krishna-Hensel, Sai Felicia (ed.) (2012) *New Energy Frontiers: Critical Energy and the Resource Challenge*. Aldershot: Ashgate.

Kulchyski, Peter and Frank Tester (2007) *Kiumajuk [Talking Back]: Game Management and Inuit Rights in Nunavut 1900 to 1970.* Vancouver: UBC Press.

Kuletz, Valerie L. (1998) *The Tainted Desert: Environmental and Social Ruin in the American West*. New York: Routledge.

Lambert, Andrew (2009) *Franklin: Tragic Hero of Polar Navigation*. London: Faber and Faber.

Lamers, Machiel, Daniela Liggett and Tina Tin (2014) 'Strategic Thinking for the Antarctic Environment: The Use of Assessment Tools in Governance'. In Tina Tin, Daniela Liggett, Patrick T. Maher and Machiel Lamers (eds) *Antarctic Futures: Human Engagement with the Antarctic Environment*. Dordrecht: Springer.

Laruelle, Marlene (2013) *Russia's Arctic Strategies and the Future of the North*. New York: M.E. Sharpe.

Laxness, Halldór (2004 [1948]) *The Atom Station*. London: Vintage Books.

Leane, Elizabeth (2012) *Antarctica in Fiction*. Cambridge: Cambridge University Press.

Lemke, Peter and Jiawen Ren et al. (2007) 'Observations: Changes in Snow, Ice and Frozen Ground'. In S. Solomon, D. Qin, M. Manning, Z. Chen, M. Marquis, K.B. Averyt, M. Tignor and H.L. Miller (eds) *Climate Change 2007: The Physical Science Basis. Contribution of Working Group I to the Fourth Assessment Report of the Intergovernmental Panel on Climate Change*. Cambridge: Cambridge University Press.

Lopez, Barry (1986) *Arctic Dreams: Imagination and Desire in a Northern Landscape*. New York: Macmillan.

Lovecraft, H.P. (2005 [1936]) *At the Mountains of Madness*. New York: The Modern Library.
Lynge, Finn (1992) *Arctic Wars: Animal Rights, Endangered Peoples*. Hanover, NH: University Press of New England.
Lythe, Matthew B. and David G. Vaughan (2001) 'BEDMAP: A New Ice Thickness and Subglacial Topographic Model of Antarctica'. *Journal of Geophysical Research* 106: 11335–51.
McCamley, Nick (2002) *Cold War Secret Nuclear Bunkers: The Passive Defence of the Western World during the Cold War*. Barnsley: Pen and Sword Books.
McCannon, John (1998) *Red Arctic: Polar Exploration and the Myth of the North in the Soviet Union 1932–1939*. Oxford: Oxford University Press.
McCannon, John (2012) *A History of the Arctic: Nature, Exploration and Exploitation*. London: Reaktion Books.
Macalister, Terry (2014) 'Greenland Explores Arctic Mineral Riches amid Fears for Pristine Region'. *The Guardian*, 5 January. Available at: http://www.theguardian.com/world/2014/jan/05/greenland-mines-arctic-fears-pristine-environment
Mackinder, Halford J. (1904) 'The Geographical Pivot of History'. *The Geographical Society* 23(4), 421–37.
Mackinder, Halford J. (1919) *Democratic Ideals and Reality: A Study in the Politics of Reconstruction*. Washington, DC: National Defense University.
McClure, Robert (2013 [1856]) *The Discovery of a Northwest Passage*. Victoria, BC: TouchWood Editions.
McCorristine, Shane (2014) 'Polar Otherworlds: Dreams and Ghosts in Arctic Exploration'. *Nimrod: The Journal of the Ernest Shackleton Autumn School* 8: 86–101.
McCreary, Tyler and Richard Milligan (2014) 'Pipelines, Permits, and Protests: Carrier Sekani Encounters with the Enbridge Northern Gateway Project'. *Cultural Geographies* 21(1): 115–29.
Mair, Charles (1908) *Through the Mackenzie Basin: A Narrative of the Athabasca and Peace River Treaty Expedition of 1899*. Toronto: William Briggs.
Marino, E. (2012) 'The Long History of Environmental Migration: Assessing Vulnerability Construction and Obstacles to Successful Relocation in Shishmaref, Alaska'. *Global Environmental Change* 22(2): 374–81.
Massey, Doreen (2005) *For Space*. London: Sage.
Menon, K.S.R. (1982) 'The Scramble for Antarctica'. *South*, 18, 11–13.

Milton Freeman Research Ltd (1976) *Report of the Inuit Land Use and Occupancy Project*, 3 vols. Ottawa: Indian and Northern Affairs Canada.

Mitchell, Timothy (2002) *Rule of Experts: Egypt, Techno-Politics, Modernity*. Berkeley, CA: University of California Press.

Morley, David and Kevin Robins (1995) 'Techno-Orientalism: Japan Panic'. In *Spaces of Identity: Global Media, Electronic Landscapes and Cultural Boundaries*. London: Routledge, pp. 147–73.

Murashko, Olga (2008) 'Protecting Indigenous Peoples' Rights to Their Natural Resources: The Case of Russia'. *Indigenous Affairs* 3–4: 48–59.

Nadasdy, Paul (2003) *Hunters and Bureaucrats: Power, Knowledge, and Aboriginal-State Relations in the Southwest Yukon*. Vancouver, BC: University of British Columbia Press.

Nally, David (2014) 'Governing Precarious Lives: Land Grabs, Geopolitics, and "Food Security"'. *The Geographical Journal*, DOI: 10.1111/geoj.12063.

Naylor, Simon, Katrina Dean and Martin Siegert (2008) 'The IGY and the Ice Sheet: Surveying Antarctica'. *Journal of Historical Geography* 34: 575–94.

Needell, Allan A. (2000) *Science, Cold War and the American State*. Washington, DC: Smithsonian Institution.

Nielsen, Kristian H., Henry Nielsen and Janet Martin-Nielsen (2014) 'City under the Ice: The Closed World of Camp Century in Cold War Culture'. *Science as Culture* 23(4): 443–64.

Nolan, Peter (2013) 'Imperial Archipelagos'. *New Left Review* 80: 77–95.

Nuttall, Mark (1992) *Arctic Homeland: Kinship, Community and Development in Northwest Greenland*. Toronto: University of Toronto Press.

Nuttall, Mark (1998) *Protecting the Arctic: Indigenous Peoples and Cultural Survival*. London: Routledge.

Nuttall, Mark (2008) 'Self-Rule in Greenland: Towards the World's First Independent Inuit State?' *Indigenous Affairs* 3–4: 64–70.

Nuttall, Mark (2009) 'Living in a World of Movement: Human Resilience to Environmental Instability in Greenland'. In Susan A. Crate and Mark Nuttall (eds) *Anthropology and Climate Change: From Encounters to Actions*. Walnut Creek, CA: Left Coast Press.

Nuttall, Mark (2010) *Pipeline Dreams: People, Environment, and the Arctic Energy Frontier*. Copenhagen: IWGIA.

Nuttall, Mark (2012a) 'Imagining and Governing the Greenlandic Resource Frontier'. *The Polar Journal* 2(1): 113–24.

Nuttall, Mark (2012b) 'Tipping points and the human world: living with change and thinking about the future'. *Ambio* 4(1): 96–105.

Nuttall, Mark (2012c) 'The Isukasia Iron Ore Mine Controversy: Extractive Industries and Public Consultation in Greenland'. *Nordia Geographical Publications* 41(5): 23–34.

Nuttall, Mark (2013) 'Zero-Tolerance, Uranium and Greenland's Mining Future'. *The Polar Journal* 3(2): 368–83.

Nuttall, Mark (2015) 'Subsurface Politics: Greenlandic Discourses on Extractive Industries'. In Leif Christian Jensen and Geir Hønneland (eds) *Handbook of the Politics of the Arctic*. Cheltenham: Edward Elgar (in press).

Nuttall, Mark and Terry V. Callaghan (eds) (2000) *The Arctic: Environment, People, Policy*. Amsterdam: Harwood.

Nuttall, Mark, Fikret Berkes, Bruce Forbes, Gary Kofinas, Tatiana Vlassova and George Wenzel (2005) 'Hunting, Herding, Fishing, and Gathering: Indigenous Peoples and Renewable Resource Use in the Arctic'. In ACIA, *Arctic Climate Impact Assessment: Scientific Report*. Cambridge: Cambridge University Press, pp. 660–702.

Ó Tuathail, Gearóid (1996) *Critical Geopolitics: The Politics of Writing Global Space*. Minneapolis, MN: University of Minnesota Press.

Ó Tuathail, Gearóid and Simon Dalby (1998) 'Re-Thinking Geopolitics: Towards a Critical Geopolitics'. In Gearóid Ó Tuathail and Simon Dalby (eds) *Rethinking Geopolitics*. New York: Routledge, pp. 1–15.

Overland, James E. and Muyin Wang (2013) 'When Will the Summer Arctic Be Nearly Sea Ice Free?'. *Geophysical Research Letters* 40: 2097–101.

Paris, Max (2013) 'Canada's Claim to Arctic Riches Includes the North Pole'. CBC News, 9 December. Available at: http://www.cbc.ca/news/politics/canada-s-claim-to-arctic-riches-includes-the-north-pole-1.2456773

Passelac-Ross, Monique and Verónica Potes (2007) 'Crown Consultation with Aboriginal Peoples in Oil Sands Development: Is it Adequate, Is it Legal?'. CIRL Occasional Paper No. 19. University of Calgary: Canadian Institute of Resources Law.

Pearson, Charles Henry (1893) *National Life and Character: A Forecast*. New York: Macmillan.

Pike, David L. (2010) 'Wall and Tunnel: The Spatial Metaphorics of Cold War Berlin'. *New German Critique* 37(2): 73–94.

Pinkerton, Alasdair and Matt Benwell (2014) 'Rethinking Popular Geopolitics in the Falklands/Malvinas Sovereignty Dispute:

Creative Diplomacy and Citizen Statecraft'. *Political Geography* 38: 12–22.

Poe, Edgar Allan (1838) *The Narrative of Arthur Gordon Pym of Nantucket*. New York: Harper & Brothers.

Pollock, Henry (2010) *A World Without Ice*. London: Avery.

Porsild, Charles (1998) *Gamblers and Dreamers: Women, Men, and Community in the Klondike*. Vancouver, BC: University of British Columbia Press.

Potter, Russell A. (2007) *Arctic Spectacles: The Frozen North in Visual Culture, 1818–1875*. Seattle, WA: University of Washington Press.

Powell, Richard (2008) 'Science, Sovereignty and Nation: Canada and the Legacy of the International Geophysical Year 1957–58'. *Journal of Historical Geography* 34(4): 618–38.

Powell, Richard (2010) 'Lines of Possession?: The Anxious Constitution of a Polar Geopolitics'. *Political Geography* 29: 74–7.

Powell, Richard and Klaus Dodds (eds) (2014) *Polar Geopolitics? Knowledges, Resources and Legal Regimes*. Cheltenham: Edward Elgar.

Price, Simon J., Jonathan R. Ford, Anthony H. Cooper and Catherine Neal (2011) 'Humans as Major Geological and Geomorphological Agents in the Anthropocene: The Significance of Artificial Ground in Great Britain'. *Philosophical Transactions of the Royal Society* 369(1938): 1056–84.

Reilly, Matthew (1998) *Ice Station*. London: Macmillan.

Revelle, Roger and Hans E. Suess (1957) 'Carbon Dioxide Exchange between Atmosphere and Ocean and the Question of an Increase of Atmospheric CO_2 During the Past Decades'. *Tellus* 9: 18–27.

Ries, Christopher Jacob (2012) 'On Frozen Ground: William E. Davies and the Military Geology of Northern Greenland 1952–1960'. *The Polar Journal* 2(2): 334–57.

Rink, Henrik (1974 [1877]) *Danish Greenland: Its People and Products*. Montreal: McGill-Queen's University Press.

Rival, Laura (2005) 'The Growth of Family Trees: Understanding Huaorani Perceptions of the Forest'. In Alexandre Surrallés and Pedro García Hierro (eds) *The Land Within: Indigenous Territory and the Perception of Environment*. Copenhagen: International Work Group for Indigenous Affairs.

Roberts, Peder (2011) *The European Antarctic: Science and Strategy in Scandinavia and the British Empire*. Basingstoke: Macmillan.

Roberts, Peder (2014) 'Scientists and Sea Ice under Surveillance in the Early Cold War'. In Simone Turchetti and Peder Roberts (eds) *The Surveillance Imperative*. Basingstoke: Palgrave, pp. 125–46.

Robinson, R.V. (1960) 'Experiment in Visual Orientation during Flights in the Antarctic'. *International Bulletin of the Soviet Antarctic Expeditions* 18: 28–9.

Rollins, James (2003) *Ice Hunt*. New York: HarperCollins.

Rosenthal, Elisabeth (2012) 'Race Is On as Ice Melt Reveals Arctic Treasures'. *The New York Times*, 18 September. Available at: http://www.nytimes.com/2012/09/19/science/earth/arctic-resources-exposed-by-warming-set-off-competition.html

Rothwell, Donald (2013) 'The Antarctic Whaling Case: Litigation in the International Court and the Role Played by NGOs'. *The Polar Journal* 3: 399–414.

Roucek, Joseph S. (1983) 'The Geopolitics of the Arctic'. *American Journal of Economics and Sociology* 42(4): 463–71.

Rozwadowski, Helen M. (2012) 'Arthur C. Clarke and the Limitations of the Ocean as a Frontier'. *Environmental History* 17(3): 578–602.

Saarinen, Aino (2009) 'A Circumpolar Case: Networking against Gender Violence across the East-West Border in the European North'. *Signs* 34(3): 519–24.

Said, Edward W. (1978) *Orientalism*. London: Penguin.

Said, Edward W. (1993) *Culture and Imperialism*. London: Vintage.

Sale, Richard and Eugene Potapov (2010) *The Scramble for the Arctic: Ownership, Exploitation and Conflict in the Far North*. London: Frances Lincoln.

Scott, Heidi V. (2008) 'Colonialism, Landscape, and the Subterranean'. *Geography Compass* 2(6): 1853–69.

Scott, James C. (1998) *Seeing Like a State: How Certain Schemes to Improve the Human Condition Have Failed*. New Haven, CT: Yale University Press.

Scott, Shirley V. (2014) 'Australia's Decision to Initiate *Whaling in the Antarctic*: Winning the Case versus Resolving the Dispute'. *Australian Journal of International Affairs* 68: 1–16.

Seaver, Kirsten A. (1996) *The Frozen Echo: Greenland and the Exploration of North America, c.A.D.1000–1500*. Stanford, CA: Stanford University Press.

Seed, Patricia (1995) *Ceremonies of Possession in Europe's Conquest of the New World, 1492–1640*. Cambridge: Cambridge University Press.

Shadian, Jessica M. (2014) *The Politics of Arctic Sovereignty: Oil, Ice, and Inuit Governance*. London: Routledge.

Shambaugh, David (2013) *China Goes Global: The Partial Power*. Oxford: Oxford University Press.

Sharma, S., J.A. Ogren, A. Jefferson, K. Eleftheriadis, E. Chan, P.K. Quinn and J.F. Burkhart (2013) 'Black Carbon in the Arctic'. *Arctic Report Card (Update for December 2013)*. Washington DC: National Oceanic and Atmospheric Administration. Available at: http://www.arctic.noaa.gov/report13/black_carbon.html

Short, John Rennie (2009) *Cartographic Encounters: Indigenous Peoples and the Exploration of the New World*. London: Reaktion Books.

Siegert, Martin J. (2000) 'Antarctic Subglacial Lakes'. *Earth Science Reviews* 50: 29–50.

Siple, Paul (1948) 'The Application of Geographical Research to US Army Needs'. *Professional Geographer* 7(1): 1–3.

Smolka, Harry Peter (1938) *Forty Thousand against the Arctic: Russia's Polar Empire*. London: Hutchinson.

Smucker, Samuel (1860) *Arctic Explorations and Discoveries during the Nineteenth Century*. New York: C.M. Saxton, Barker and Co.

Soikan, Meitiaki Ole (2009) 'The Social, Environmental and Cultural Effects of Extractive Industries in Kajiado District, Rift Valley Province, Kenya: A Case Study of (Gypsum and Limestone) Cement Factories and Soda Ash Companies'. Presentation at the International Conference on Extractive Industries and Indigenous Peoples, Manila, Philippines, 23–25 March.

Spufford, Francis (1997) *I May Be Some Time: Ice and the English Imagination*. London: Palgrave.

Stammler, Florian and Bruce C. Forbes (2006) 'Oil and Gas Development in Western Siberia and Timan-Pechora'. *Indigenous Affairs* 2–3/06, 48–57.

Stefansson, Vijhjalmur (1921) *The Friendly Arctic: The Story of Five Years in Polar Regions*. New York: Macmillan.

Stefansson, Vijhjalmur (1922) *The Northward Course of Empire*. London: George G. Harrap.

Steinberg, Philip (2001) *The Social Construction of the Ocean*. Cambridge: Cambridge University Press.

Steinberg, Philip E. (2010) 'The Deepwater Horizon, The Mavi Marmara, and the Dynamic Zonation of Ocean-Space'. *The Geographical Journal* 177: 12–16.

Steinberg, Philip (2013) 'Of Other Seas: Metaphors and Materialities in Maritime Regions'. *Atlantic Studies: Global Currents* 10(2): 156–69.

Steinberg, Philip E. (2014) 'Steering between Scylla and Charybdis: The Northwest Passage as Territorial Sea'. *Ocean Development & International Law* 45: 84–106.

Stewart, E. and D. Draper (2008) 'The Sinking of the MS Explorer'. *Arctic* 61(2): 224–8.

Struzik, Ed (2013) 'China's New Arctic Presence Signals Future Development'. *Yale Environment 360*, 4 June. Available at: http://e360.yale.edu/feature/chinas_new_arctic_presence_signals_future_development/2658/

Stuhl, Andrew (2013) 'The Politics of the "New North": Putting History and Geography at Stake in Arctic Futures'. *The Polar Journal* 3(1): 94–119.

Tester, Frank and Peter Kulchyski (1994) *Tammarniit (Mistakes): Inuit Relocation in the Eastern Arctic, 1939–63*. Vancouver, BC: University of British Columbia Press.

Tsing, Anna (2003) 'Natural Resources and Capitalist Frontiers'. *Economic and Political Weekly* 38(48): 5100–6.

Turchetti, Simone and Peder Roberts (eds.) (2014) *The Surveillance Imperative: Geosciences during the Cold War and Beyond*. Basingstoke: Macmillan.

Verne, Jules (1889) *The Purchase of the North Pole*.

Virilio, Paul (1989) *War and Cinema: The Logistics of Perception*. London: Verso.

Virilio, Paul (2002) *Desert Screen: War at the Speed of Light*. London: Continuum.

Wadhams, Peter (2012) 'Arctic Ice Cover, Ice Thickness and Tipping Points'. *Ambio* 41(1): 23–33.

Wager, Walter H. (1962) *Camp Century: City under the Ice*. Philadelphia, PA: Chilton Books.

Wassman, Paul and Timothy Lenton (2012) 'Arctic Tipping Points in an Earth System Perspective'. *Ambio* 41(1): 1–9.

Watt-Cloutier, Sheila (2005) 'Petition to the Inter American Commission on Human Rights Seeking Relief from Violations Resulting from Global Warming Caused by Acts and Omissions of the United States'. Available at: http://earthjustice.org/sites/default/files/library/legal_docs/summary-of-inuit-petition-to-inter-american-council-on-human-rights.pdf

Watts, Michael (2001) 'Petro-Violence: Community, Extraction, and Political Ecology of a Mythic Commodity'. In Nancy Lee Peluso and Michael Watts (eds) *Violent Environments*. Ithaca, NY: Cornell University Press, pp.189–212.

Watts, Michael (2012) 'A Tale of Two Gulfs: Life, Death, and Dispossession along Two Oil Frontiers'. *American Quarterly* 64(3): 437–67.

Watts, Michael, Arthur Mason and Hannah Appel (eds) (2014) *Oil

Talk: The Secret Lives of the Oil and Gas Industry. Ithaca, NY: Cornell University Press.

Weiss, Erik D. (2001) 'Cold War under the Ice: The Army's Bid for a Long-Range Nuclear Role, 1959–1963'. *Journal of Cold War Studies* 3(3): 31–58.

Weizman, Eyal (2003) *A Civilian Occupation*. London: Verso.

Wenzel, George W. (1991) *Animal Rights, Human Rights: Ecology, Economy and Ideology in the Canadian Arctic*. Toronto: University of Toronto Press.

Williams, Alison (2009) 'A Crisis in Aerial Sovereignty? Considering the Implications of Recent Military Violations of National Airspace'. *Area* 42(1): 51–9.

Williams, Glyn (2003) *Voyages of Delusion: The Search for the Northwest Passage in the Age of Reason*. New Haven, CT: Yale University Press.

Williams, Glyn (2009) *Arctic Labyrinth: The Quest for the Northwest Passage*. Toronto: Viking Canada.

Wright, Shelley (2014) *Our Ice is Vanishing*. Montreal: McGill-Queens University Press.

Xinhua News Agency (2011) 'Greenland sees big potential in China's Arctic presence', 27 November.

Yusoff, Kathryn, Elizabeth Grosz, Nigel Clark, Arun Saldanha and Catherine Nash (2012) 'Geopower: A Panel on Elizabeth Grosz's "Chaos, Territory, Art: Deleuze and the Framing of the Earth"'. *Environment and Planning D: Society and Space* 30(6): 971–88.

INDEX

208

ice, 18, 22, 23, 37, 45, 55, 58–61, 73,
154
ice curtain, 58
ice sheet, 6, 16, 50, 54, 64, 75–8,
78, 86, 115, 175
pack ice, 60, 108, 152
popular representations, 58–60,
64–5
sea ice, 4, 16, 19, 23, 50–1, 54–5,
68, 71, 73, 88, 117, 157–8, 159,
176–8
subglacial lakes, 64, 76, 77–8
Iceland, 33, 88, 107–8, 145, 146, 160,
168, 170
India, 91, 162–4, 182
indigenous peoples, 11–13, 27, 29,
83, 95–101, 186
homeland, 95, 98, 101, 112, 120,
127, 129–30, 150
hunting, 12, 29, 48, 96
knowledge, 13, 53, 98, 113, 153,
183
land claims, 12, 13, 24, 46, 83, 98,
102–3, 128–9, 151
pollution, 14, 83–5, 103–5, 126
resources, 98–9, 100, 111–13,
126–8, 129–30
self determination, 47, 101
Self-Rule, 47, 131
see also Inuit
infrastructure, 26, 33, 38, 63, 66,
107, 115, 125–6, 140, 149, 161,
171, 173
International Court of Justice (ICJ),
39, 94
International Geophysical Year (IGY),
72–5, 76, 79, 89, 91, 137
International Hydrographic
Organization (IHO), 8–9
International Panel on Climate
Change (IPCC), 51–2, 54
International Polar Year (IPY), 69
Inuit, 6, 34, 48–9, 53, 65, 98, 105–6,
113, 176, 178, 183–4, 186
Declaration on Sovereignty, 112, 127

Inuit Circumpolar Council (ICC),
111–12, 127, 183

Japan, 38–9, 68, 91, 94–5, 115, 138,
145, 151, 166, 168

land claims, see indigenous
peoples
Landseer, Edwin, 10
legalization, 41–6
Lego, 49, 175

Mackinder, Halford, 3, 147–8,
151–2
Madrid Protocol, see Protocol on
Environmental Protection
Malaysia, 137, 164
maps/mapping, 11, 18, 23, 35–6, 43,
61–2, 70, 72, 76, 82, 97, 98, 150,
155, 174
marine protected areas (MPAs), 21,
94, 138, 166, 180–2
Massey, Doreen, 31–2
mineral exploitation, 20, 37, 49,
83–5, 137–8
mining, 2, 37, 48, 83, 85, 115–17,
131–6, 138, 140, 145, 149–50,
172
mobility, 23, 37, 68, 114, 151, 153–8,
169, 177, 185
MV Explorer, 14

NATO, 108
New Zealand, 9, 23, 77, 82, 89, 142,
162, 180, 182
Non-Aligned Movement, 163–4
North Atlantic, 33–4, 71, 105,
107
North Pole, 1, 4, 42, 55, 70, 80, 116,
151–2, 176
Northeast Passage, 1, 121
Northern Sea Route (NSR), 40, 45,
154–8
Northwest Passage, 1–2, 17, 45, 59,
106, 121–2, 140, 155–7